PRAISE FOR
DR. ART'S GUIDE TO PLANET EARTH

"This is an outstanding book. Vividly, clearly and concisely Art Sussman explains how our planet works and what can happen when the balance of nature is upset. It will capture the imagination of readers of all ages and invoke a sense of wonder. I absolutely recommend *Dr. Art's Guide to Planet Earth* – it deserves a place not only in every classroom but also every home. Get it now!"

Dr. Jane Goodall
Ecologist, Author

"I recommend this highly readable book for people of all ages who are interested in learning how the earth's physical and life systems are interconnected."

Dr. Bruce Alberts
President, National Academy of Sciences

"What a fun read. This book artfully explores Earth's systems while also engaging the reader's imagination. Complex ideas are presented in a way that is truly understandable! This isn't just for the kids; adults will find themselves reading the book to answer their own questions about how the Earth works."

Dr. Bora Simmons
Director, National Project for Excellence in Environmental Education

"As Project Director of Public Understanding of Science and Technology, I am always looking for materials that effectively communicate science. *Dr. Art's Guide to Planet Earth* is an excellent resource for the general public who want to understand the environmental issues that they encounter in the media, and in the choices that they make as citizens and consumers. The writing is enjoyable and the graphics are easy to understand."

Judy Kass
American Association for the Advancement of Science

"This very good book is friendly, well thought out and contains interesting information. It also takes a unique point of view that could be very useful for learning about systems and interactions."

Dr. Goery Delacote

Executive Director, Exploratorium, San Francisco

"This is a superb primer on the web of life. My hope is that every child who reads it entice their parents to read it too, in hopes that adults will join the children in understanding we have only one precious web of life on Earth."

Paul Hawken

Author, Chairman, The Natural Step, US
Founder, Smith and Hawken

"What a wonderfully concise and brilliantly simple presentation! As an environmental science educator for 20 years, I'm glad to have such a user-friendly teaching tool. Highly recommended for all Earthlings!"

Carol Fialkowski

Field Museum of Natural History, Chicago

"I have used Dr. Art's ideas to teach teachers and to teach high school students. Both groups love his approach. They develop a deep understanding of how our planet works, and they enjoy themselves as they learn."

Teresa Hislop

Utah Mentor Science Teacher

DR. ART'S GUIDE TO PLANET EARTH

FOR EARTHLINGS AGES 12 TO 120

Art Sussman, Ph.D.

Chelsea Green Publishing Company • White River Junction, VT
WestEd • San Francisco, CA

Dr. Art's Guide to Planet Earth

Design: Susan M. Young
Illustration: Emiko-Rose Koike

Library of Congress Cataloging-in-Publication data is available.
Please contact Chelsea Green Publishing Company at 1-800-639-4099.

Photos courtesy of:
Artville: pp. 36, 100; Corbis Images: pp. 80, 81; Eyewire: cover, pp. i, 5, 6, 9, 10, 15, 26, 41, 43, 44, 53, 60, 71, 75, 77, 79, 83, 88, 96, 99, 102, 103, 104, 106, 109, 114; Masterphotos; pp. 1, 8, 11, 13, 14, 15, 17, 18, 19, 20, 24, 31, 33, 43, 46, 47, 50, 51, 75; NASA: pp. 1, 2, 3, 17, 78, 90; Photodisc; cover, pp. i, iii, 5, 19, 21, 29, 34, 35, 45, 54, 55, 60, 63, 66, 68, 74, 75, 77, 79, 80, 83, 84, 87, 99, 102, 111, 115.

ISBN: 1-890132-73-X

Chapter 1 - Introducing Planet Earth

Earth Is Whole ____ 2
Systems within Systems within Systems ____ 4
The Earth System ____ 8
Earth's Matter ____ 10
Earth's Energy ____ 12
Earth's Life ____ 14
Three Principles ____ 16
Book Overview ____ 17

Chapter 2 - Matter Cycles

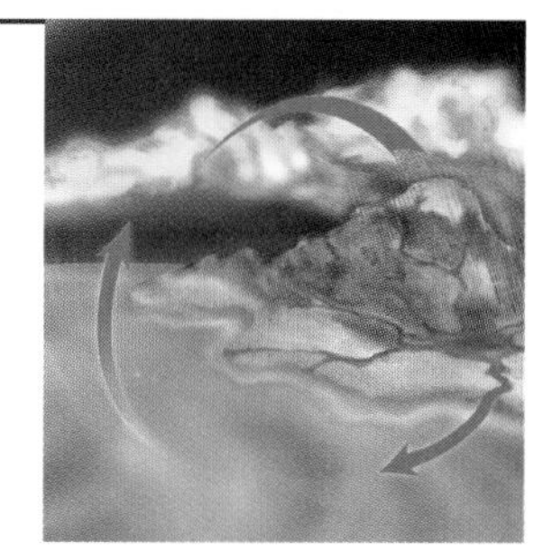

Earth's Solid Stuff ____ 20
The Rock Cycle ____ 24
Earth's Liquid Stuff ____ 26
The Water Cycle ____ 28
Earth's Gas Stuff ____ 34
The Carbon Cycle ____ 36
Today's Carbon Cycle ____ 40

Chapter 3 - Energy Flows

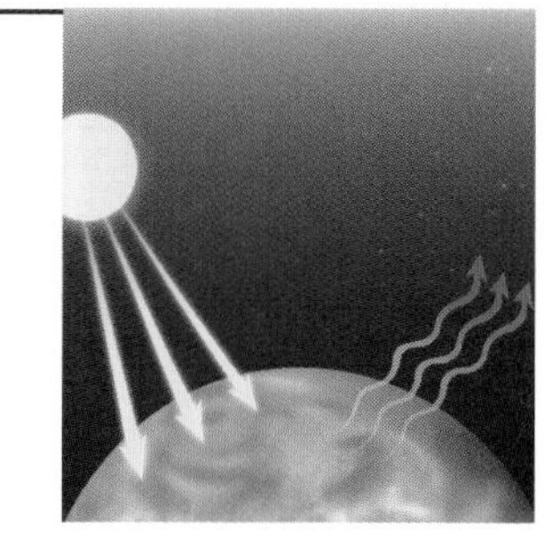

Earth's Energy ____ 44
Part of a Larger System ____ 46
Energy from the Sun ____ 48
The Greenhouse Effect ____ 50
Earth's Internal Energy ____ 54
Earth's Energy Budget ____ 56

CHAPTER 4 - Life Webs

A Living Planet 60
Watching Earth Breathe 62
Who is in the Web? 66
Ecosystems and Feedback Loops 68
Shredding the Web 74

CHAPTER 5 - Think Globally

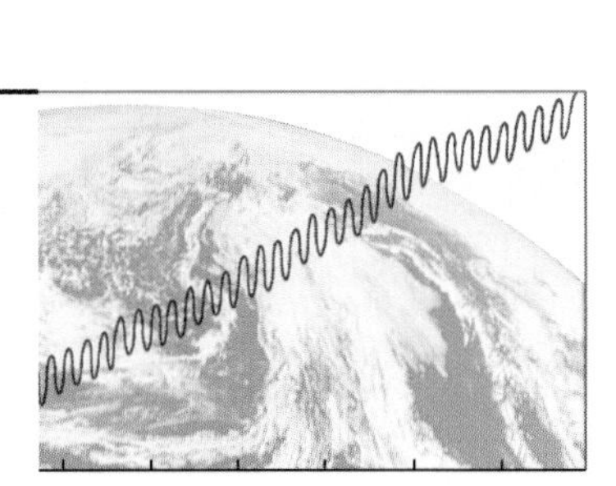

Save the Planet? 78
Extinction 80
The Ozone Layer 86
Climate Change 92
Ice Age or Hot House? 96

CHAPTER 6 - Act Locally

Healthy Air, Water & Food 100
The Three R's 102
Local Ecosystems 104
What About Energy? 106
What Can I Do? 108
Making a Difference 112
Not the End 116

chapter 1

introducing planet earth

Earth Is Whole

Systems within Systems within Systems

The Earth System

Earth's Matter

Earth's Energy

Earth's Life

Three Principles

Book Overview

earth is whole

One of humanity's major discoveries is that we live on a round planet. We laugh about the idea that Earth is flat. Yet, we ourselves are in the midst of an even greater change in how to understand our planet. And most of us don't know about it.

BIG IDEA

"Earth is Whole" means that all the planet's physical features and living organisms are interconnected.

When we realized that Earth is round, we learned how the places on our planet are physically connected to each other. We discovered that if we kept traveling in one direction, we would not fall off the edge. Instead we could go in a circle and return to our starting place. That was an important discovery for our ancestors.

Now we are learning something much more important than how the places on our planet are physically connected. We are discovering how Earth works as a whole system. Earth is not flat. Earth is much more than round. Earth is whole.

"Earth is Whole" means that all the planet's physical features and living organisms are interconnected. They work together in important and meaningful ways. The clouds, oceans, mountains, volcanoes, plants, bacteria and animals all play important roles in determining how our planet works.

Scientists have established a new field of science called Earth systems science to study and discover how all these parts work together. Earth systems scientists combine the tools and ideas from many scientific disciplines including geology, biology, chemistry, physics and computer science. In addition, they use modern technologies to measure key features of our planet, such as the concentration of gases in the atmosphere and the temperature of the ocean in many locations. Satellites orbiting our planet provide enormous amounts of data that Earth systems scientists use to try to understand how our planet works and what kinds of changes are happening.

Of course, human beings do more than study and measure planet Earth. Just like any other organism, we are a part of this whole Earth system. More importantly, we now have a very challenging new role to play. For the first time in our history, we can dramatically change the way the planet works as a whole. There are so many of us and we have such powerful technologies that we can change Earth's climate, destroy its ozone shield and dramatically alter the number and kinds of other organisms that share the planet with us.

Over the past five years, I have developed a method of explaining Earth systems science to my family, friends, co-workers, teachers and students. I also perform a show where I use scientific demonstrations and audience participation to introduce the three Earth systems principles that are featured in this book. My experience is that people enjoy learning about Earth systems science, and they feel they get a better perspective on how our planet works and what they can do as local and global citizens.

Dr. Art's Guide to Planet Earth helps answer one of the most important questions of the twenty-first century: Can all of us live well on our planet without damaging the whole Earth system? To answer this question, we need to understand how our planet works. That sounds much more complicated than discovering that Earth is round. Fortunately, Earth systems science can explain many of the most important features of how our planet works.

systems within systems within systems

The first step toward understanding how Earth works is to think about our planet as a system. We use the word "system" when we want to describe something that is made up of different kinds of parts that join together to form an interconnected whole. Learning to think in terms of systems is very useful because we are surrounded by all sorts of systems. In fact, each of us is our own little system.

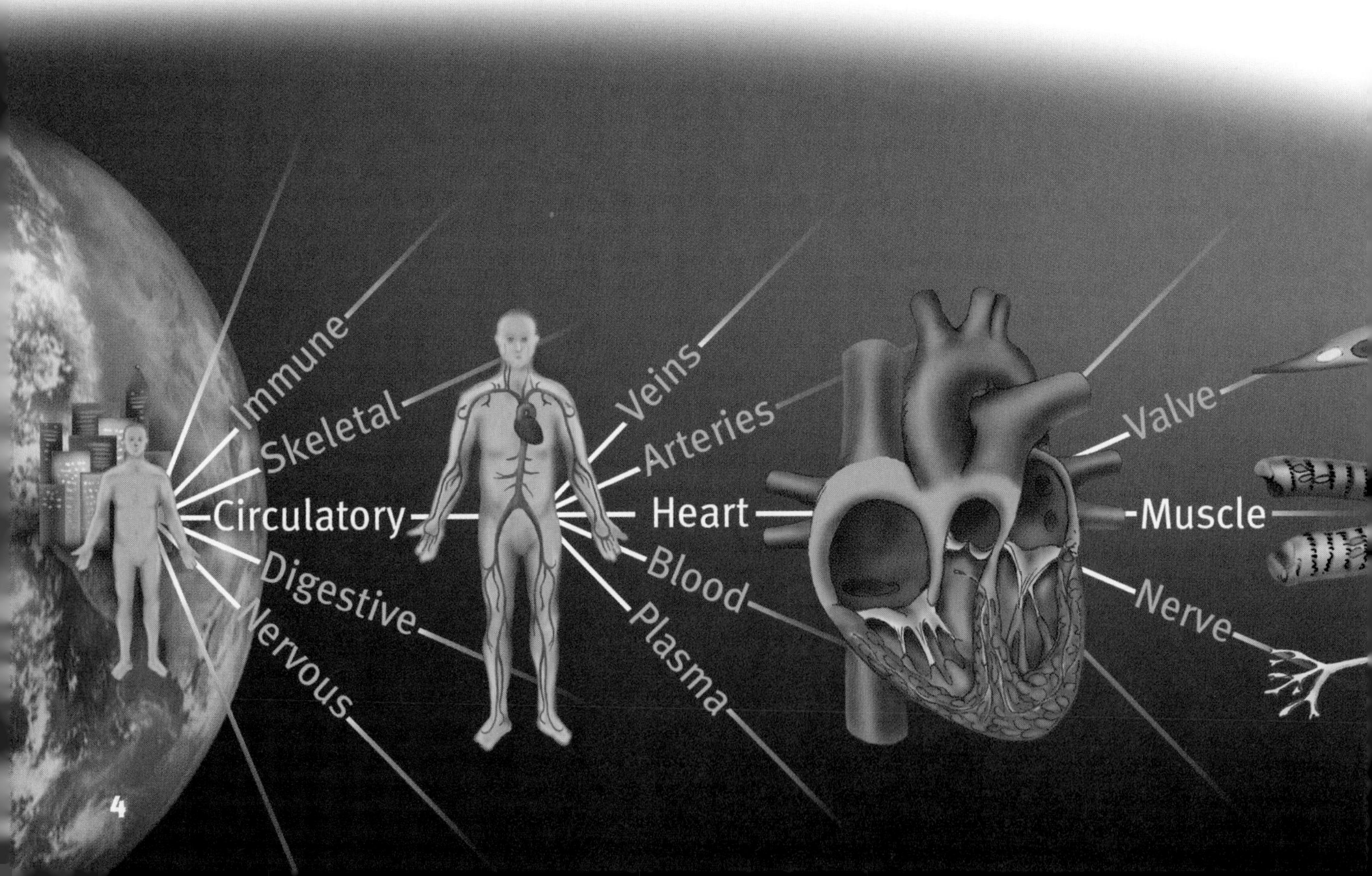

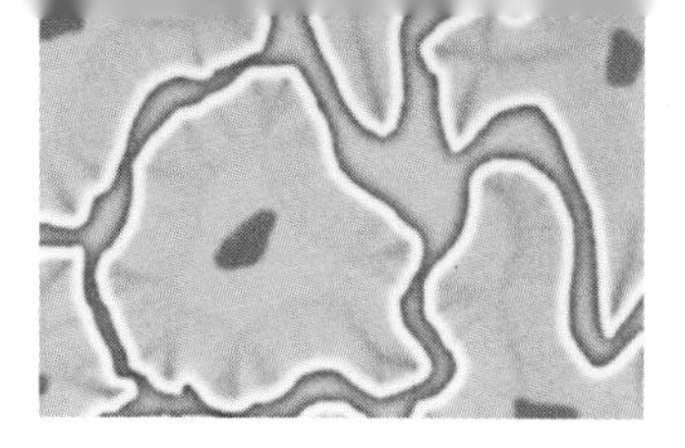

Each of us is made up of more than 200 kinds of cells. These nerve, skin, muscle, bone, red blood and gland cells all join together to form an incredible system – an individual human person. All the structures that these cells form – our skin, muscles, bones, blood vessels, internal organs – function as an interconnected whole.

Looking at ourselves as a system reveals two important features of systems:

- each part of a system can itself be described as a system;
- a system can be very different from its parts.

Each part of a system can itself be described as a system. You are a system. One of the parts of the "you system" is the way blood moves throughout your body – in other words, your circulatory system. This circulatory system is part of the bigger "you system" but it itself is a system with many parts.

The parts of the circulatory system include heart, veins, arteries and blood cells. The heart, a part of the circulatory system, is also a system made of parts. Its parts include muscle cells, nerve cells and valves. A heart muscle cell is part of the heart system but it is also a system that is made up of a cell membrane, cell nucleus and many different proteins.

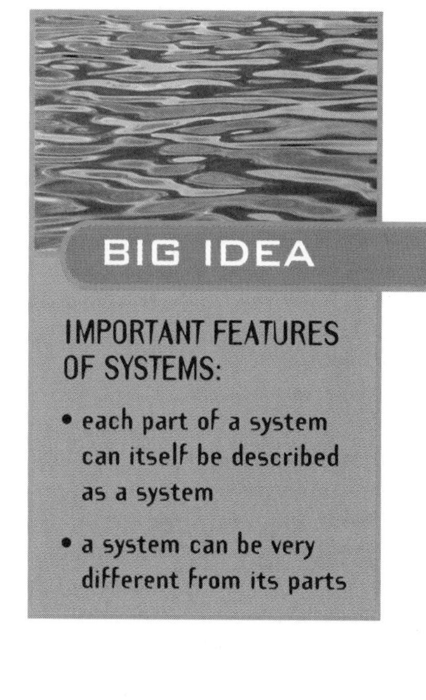

BIG IDEA

IMPORTANT FEATURES OF SYSTEMS:

- each part of a system can itself be described as a system
- a system can be very different from its parts

You could get dizzy visualizing all these systems within systems within systems that are inside each one of us. And the story does not end with us. We are not the biggest system around. Each of us systems is, in turn, part of many larger systems. Each of us is part of a family system. Each of us is part of an ecosystem. Each of us is part of an entire human system that is part of the system of life on this planet.

Why should we care about all these systems within systems within systems? The second system feature that we mentioned earlier provides an important clue.

A system can be very different from its parts. Think about your arteries, red blood cells, stomach and toenails. Your stomach is a part of who you are, but you are much more than your stomach. You are much more than the sum of your parts. As a functioning, interconnected whole, you have characteristics that do not exist in any of your parts. You have properties that transcend, that go far beyond, the qualities of your parts.

A car provides another example of a system. A car has brakes, wheels, cylinders, battery, windshield wipers, carburetor, gas tank, metal frame, steering wheel, and hundreds of other parts. Individually none of those parts will move you from your home to school, work, a restaurant or a lake. Joined together as an interconnected whole, the car system can take you away. It has properties that are qualitatively different than the properties of its parts. No part of a car gets 35 miles per gallon on the highway. No part of a car has the ability to transport you up a mountain road. Only the car as a functioning whole system has these properties.

The popular saying "the whole is more than the sum of its parts" describes this second system feature. This popular saying is much deeper than it might first appear. When we say that the whole is more than the sum of its parts, we mean that the whole system has qualities that are different than those of the parts. The whole is qualitatively different, which is a much more important difference than a mere increase in quantity.

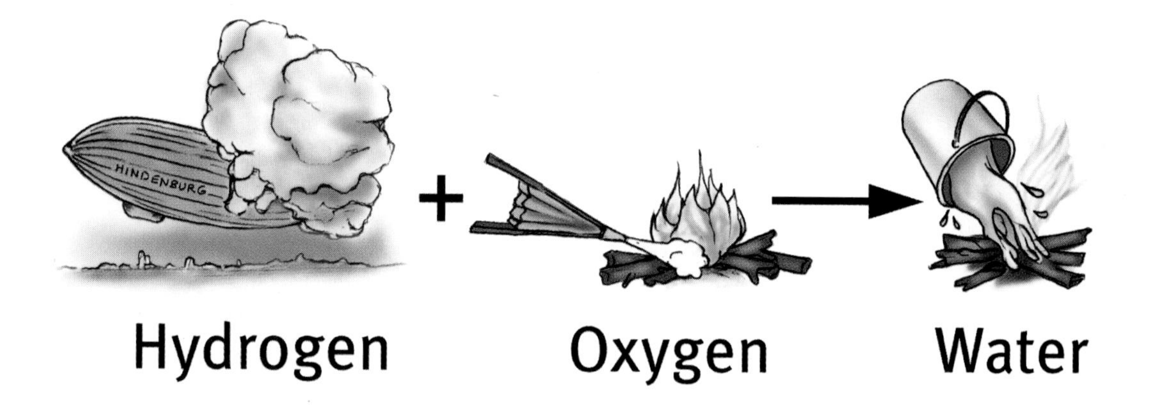

Take water as another example of a system. Water is made of hydrogen and oxygen. At normal temperatures and pressures, they are both gases. Hydrogen is highly explosive, and fires require oxygen. Put them together and you have a liquid that *extinguishes* fires. The system of hydrogen joined with oxygen (H_2O) has properties that are qualitatively very different from the parts hydrogen alone or oxygen alone.

the earth system

Many of us feel overwhelmed by the environmental issues that we encounter in newspapers and magazines, or on television, radio or the Internet. We see weird combinations of letters like PCB and CFC. We read statements from opposing Ph.D. experts, one of whom says that global warming is a serious problem while the other tells us we have nothing to worry about. How can we understand these complicated environmental issues?

BIG IDEA

THE THREE SYSTEM QUESTIONS:

- What are the parts of the system?
- How does the system function as a whole?
- How is the system itself part of larger systems?

The reason to care about "systems within systems within systems" is that systems thinking provides us with a way to understand any particular system, especially complicated ones like planet Earth. The system could be a person or your circulatory system or a car or planet Earth. No matter what the system is, we can always understand it better by asking three systems questions.

- What are the parts of the system?
- How does the system function as a whole?
- How is the system itself part of larger systems?

Dr. Art's Guide to Planet Earth uses systems thinking to help us understand how our planet works and how we can support the way our planet currently operates. We will learn three guiding principles that can provide a framework for our thinking. These principles help us focus on major concepts rather than becoming lost in confusing details. They provide a framework to guide our actions as individuals, local communities, nations and a global species. We begin with the first systems question: What are the parts of the Earth system? To understand how our planet works, I believe it is best to describe the Earth system in terms of the following three parts:

- Earth's matter
- Earth's energy
- Earth's life

In examining Earth as a whole, we are going to focus on Earth's matter, Earth's energy and Earth's life. In other words, we are going to examine from a systems point of view the stuff (matter) that exists on planet Earth, the energy that makes things happen on planet Earth, and the organisms that make our planet unique in the solar system.

earth's matter

Our planet has been circling the sun for more than four billion years. During those billions of years, the matter on our planet keeps changing its form. Water evaporates from the ocean, forms clouds and falls as snow and rain. Rocks get broken down into dirt that is washed as sediment into rivers. Plants take carbon dioxide gas from the atmosphere and convert it into solid sugars and starches. Why doesn't all the ocean water turn into mountain snow, or all the rocks turn into sediment or all the atmospheric carbon dioxide turn into sugar?

Earth still has oceans, mountains and atmospheric carbon dioxide because they are part of cycles – the water cycle, the rock cycle and the carbon cycle. Water flows in rivers back to the oceans; buried sediments reach the surface again through volcanoes; and animals chemically change sugars into carbon dioxide that goes back into the atmosphere.

Earth is a recycling planet. Essentially all the matter on Earth has been here since the planet was formed. We don't get new matter; old matter does not go away into outer space. The same matter keeps getting used over and over again. From a systems point of view, we say that Earth is essentially a closed system with respect to matter.*

* Of course, Earth is not a totally closed system with respect to matter. For example, we are constantly bombarded by meteorites. Yet, the total amount of matter that has entered the Earth system during the past three billion years is less than 0.00001% of Earth's total mass.

MATTER CYCLES

Each of the elements that is vital for life exists on Earth in a closed loop of cyclical changes. From a systems point of view, Earth is essentially a closed system with respect to matter.

earth's energy

Imagine what would happen if the sun stopped shining! This disastrous scenario emphasizes the crucial role of solar energy. Our planet relies on a constant input of energy from the sun. Earth receives an inflow of solar energy that is more than 15,000 times the amount of energy consumed by all human societies. This constant flow of solar energy into the planet system provides virtually all the energy to keep our planet warm, drive the cycles of matter and sustain life.

If Earth retained all that energy, it would become so hot that it would just boil away. But energy does not stay in any one place. Energy flows away from Earth in the form of heat radiating to outer space. The amount of energy radiating to outer space balances the amount of energy flowing in from the sun.

Note the difference between Earth's matter and Earth's energy. With respect to matter, Earth is a closed system. Matter does not enter or leave. With respect to energy, Earth is an open system. Sunlight energy flows in and heat energy escapes.

ENERGY FLOWS

The functioning of our planet relies on a constant input of energy from the sun. This energy leaves Earth in the form of heat flowing to outer space. From a systems point of view, Earth is an open system with respect to energy.

earth's life

Earth's organisms form an intricate web of interconnections, with every organism depending on and significantly affecting many others. As one very important example, virtually all communities of organisms ultimately depend on plants. Plants capture energy from the sun and store it as chemical energy. Plants are Earth's producers.

With respect to food energy, the rest of the organisms are consumers. Some eat plants, others eat animals that eat plants and some eat both plants and animals. The plants, in turn, rely on animals for pollination or for spreading seeds, and on decomposers for creating rich soil from dead waste.

With respect to life, Earth is a networked system. Not only do organisms form an interconnected web, they also participate actively in Earth's matter cycles and energy flows. Human beings depend on the web of life for the air that we breathe and the food that we eat. As our numbers have exponentially increased and our technologies have altered virtually every part of the globe, we have become a very important part of this web of life.

LIFE WEBS

A vast and intricate network of relationships connects all Earth's organisms with each other and with the cycles of matter and the flows of energy. From a systems point of view, Earth is a networked system with respect to life.

three principles

We began investigating the Earth system by asking the first system question: What are the parts of the Earth system? Our answer focused on three parts – Earth's matter, Earth's energy and Earth's life. The second system question asks: How does the system function as a whole?

Guess what? We have already answered that question. When we looked at each part of the Earth system, we learned how that part works for the planet as a whole. As a result, we can say that there are three simple principles that work together:

MATTER CYCLES

Each of the elements that is vital for life exists on Earth in a closed loop of cyclical changes. From a systems point of view, Earth is essentially a closed system with respect to matter.

ENERGY FLOWS

The functioning of our planet relies on a constant input of energy from the sun. This energy leaves Earth in the form of heat flowing to outer space. From a systems point of view, Earth is an open system with respect to energy.

LIFE

LIFE WEBS

A vast and intricate network of relationships connects all Earth's organisms with each other and with the cycles of matter and the flows of energy. From a systems point of view, Earth is a networked system with respect to life.

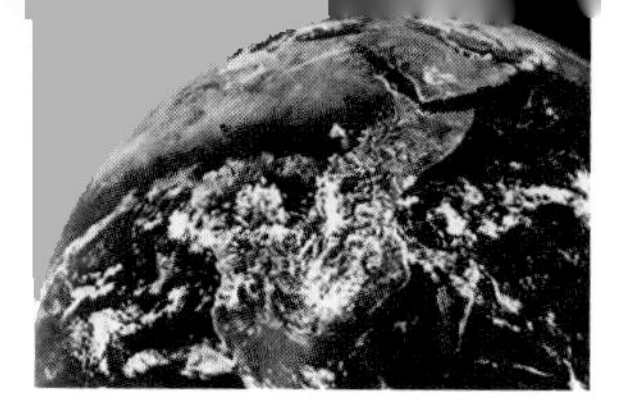

These three principles can help us understand essentially all environmental issues. When we confront an environmental issue, we should first explore the roles of matter, energy and living organisms. Where does the matter (carbon, water, pollutant) come from and where does it go? Does the issue involve changes to our planet's energy flows? How do plants, animals and microorganisms influence the issue and how are they affected by it? As a result of answering these kinds of questions, we will discover that the three guiding principles provide an organizing framework that makes common sense out of complicated issues.

Book Overview

In the twenty-first century, we find ourselves in a new world. Without meaning to, we can change the way that our planet works. At the same time, we are developing a much deeper understanding of the Earth system.

This chapter has introduced three principles that can help us focus on major concepts rather than become lost in confusing details. The next three chapters will help us understand each of these three principles much more deeply. We will be able to explain how Earth works in terms of cycles of matter, flows of energy and the web of life.

But merely knowing something, even something as important as how our planet works, is not enough. We need to apply that information in our lives. The last two chapters apply this understanding of the whole Earth system to environmental issues that we face globally and locally. The three Earth systems principles help us understand these issues, and they also provide guidance regarding what we need to do as a global species, nations, local communities and as individuals.

www.planetguide.net

matter cycles

Earth's Solid Stuff

The Rock Cycle

Earth's Liquid Stuff

The Water Cycle

Earth's Gas Stuff

The Carbon Cycle

Today's Carbon Cycle

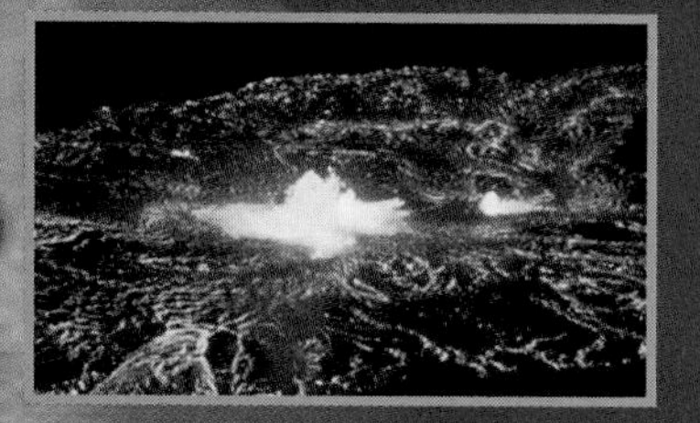

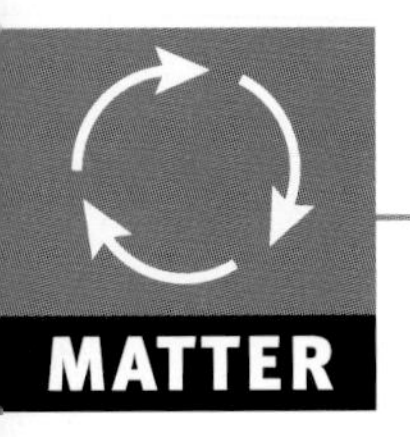

earth's solid stuff

In this chapter, we are going to explore the SYSTEM of Earth's matter. As a systems thinker, you probably began asking yourself a systems question, such as: What are the parts of the Earth matter system?

We can think of Earth's matter system as being made of three kinds of parts – solid stuff, liquid stuff and gas stuff. Scientists call these parts of the Earth matter system its geosphere (solid), hydrosphere (water) and atmosphere (gas). We begin with the geosphere, Earth's solid stuff.

We can hardly imagine the conditions 4,500,000,000 (four billion five hundred million) years ago when Earth began to take shape. As a young, new planet, Earth was an exploding ball of molten rock and metal. Material kept crashing into and sticking to the planet, causing it to grow larger and larger. As it eventually stabilized in size and cooled, the densest material settled to the center forming an iron core that causes Earth's magnetic field.

We live on a thin crust of the less dense material. This crust remained floating on the top and solidified as it cooled. If we represent our planet as a globe that is four feet in diameter, the crust would occupy just the top one quarter inch.

Most of the geosphere is very different than the solid Earth that we experience every day. Below us lies an almost completely unexplored world of extremely hot rock and metal. This material, existing in conditions of very high temperatures and pressures, melts and flows, descending thousands of miles below our feet, homes, oceans and forests. Earthquakes, volcanoes and geysers indicate the high temperatures and pressures that exist in Earth's pressure cooker interior.

Scientists thought today's continents and oceans had been the same for billions of years. In the 1960s, they found convincing evidence that challenged this view of the Earth. Their measurements, theories and data analysis changed the way we understand our planet and caused a revolution in the Earth sciences.

This revolution taught us that Earth's surface consists of about a dozen huge plates that move into, away from, over, under, and next to each other. These plates float on top of a moving layer of hotter, more fluid material. The oceans and continents are contained as parts of these plates and move with them. So, instead of staying the same for billions of years, the continents and oceans keep changing their size and location.

BIG IDEA

Earthquakes, volcanoes and geysers indicate the high temperatures and pressures that exist in Earth's pressure cooker interior.

Look how quickly they change! Only last month (okay, 225 million years ago - but that is last month on the geologic time scale), all the land mass was joined together as one huge supercontinent. By the time of the Jurassic Age (135 million years ago), some separation had occurred but Africa was still practically joined to South America. Only in the last 135 million years (less than 5% of Earth's existence) has the mighty Atlantic Ocean formed between the Americas and Africa/Europe.

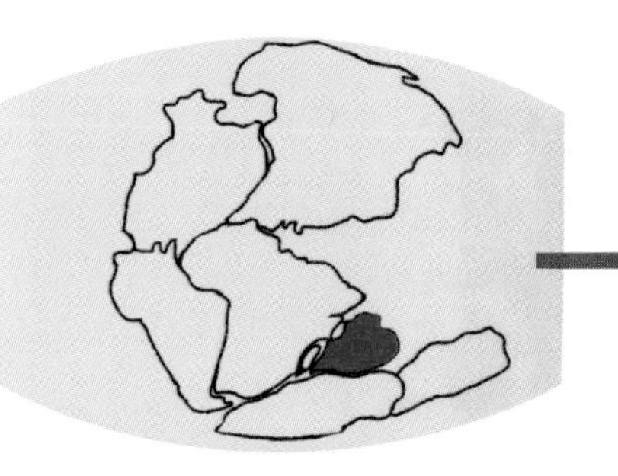

225 million years ago

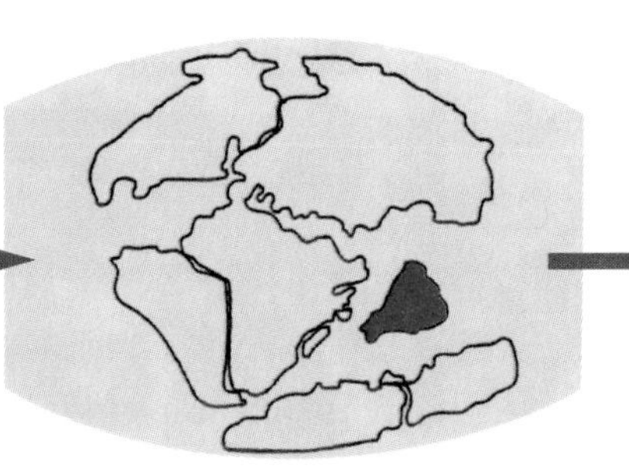

135 million years ago

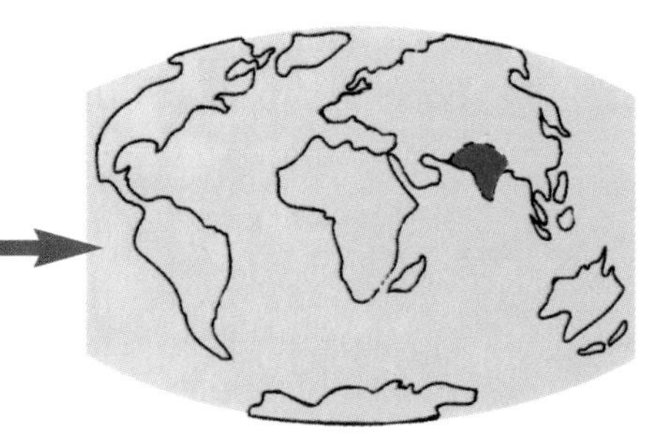

Today

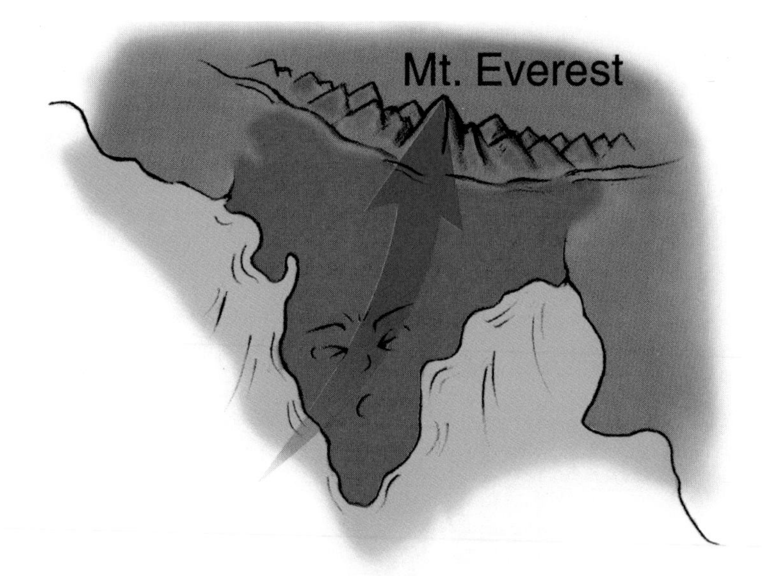

India provides another dramatic example of these changes. The current Indian landmass once existed south of the Equator, near the current location of Australia. During these hundreds of millions of years, the plate carrying today's India moved about 4,000 miles northward. As a result, India crashed into Asia approximately 40 million years ago and joined that continent. The surface crust where India and Asia collide crumpled upwards, forming the Himalayas, which includes Mt. Everest and the other nine highest peaks in the world.

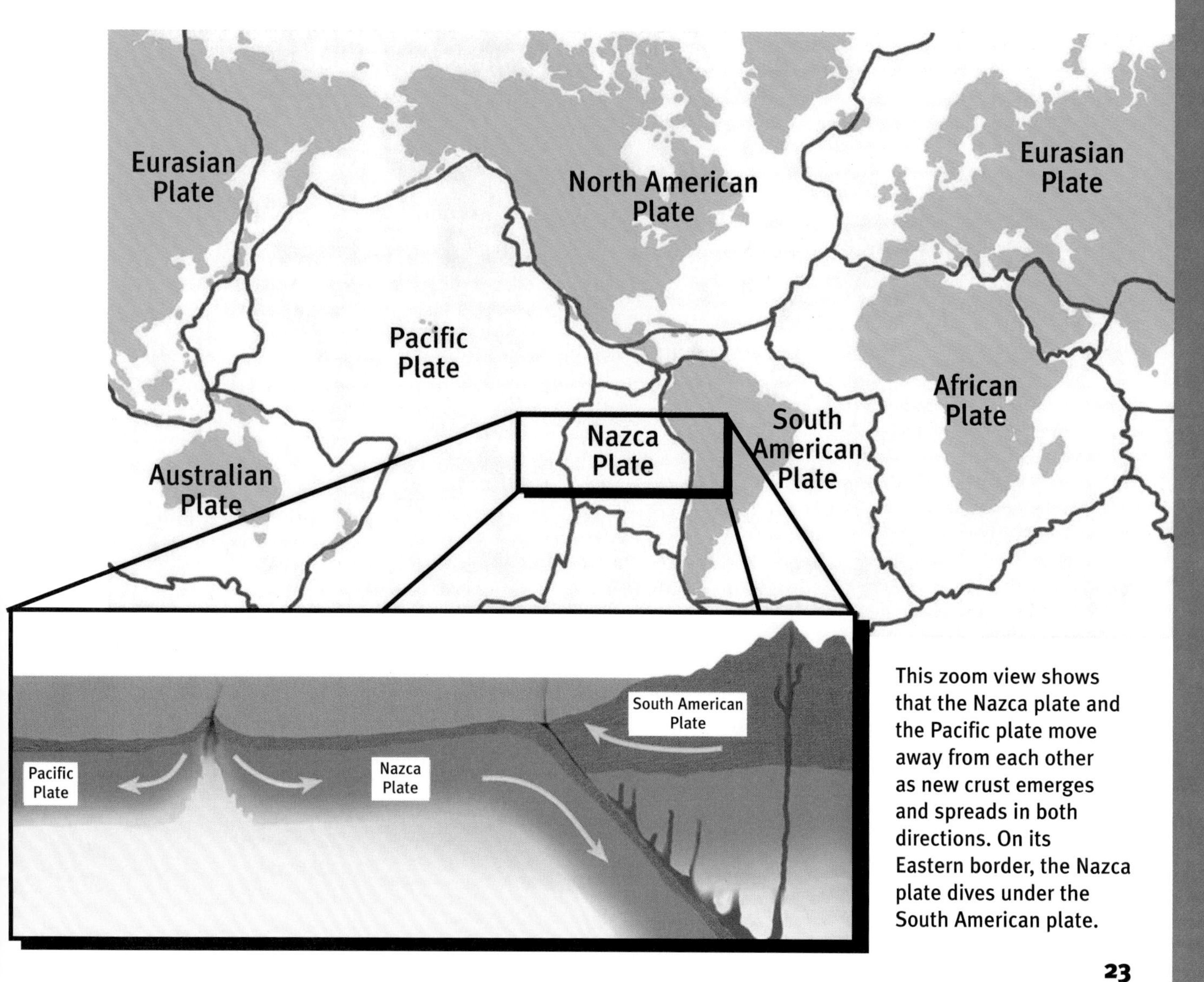

This zoom view shows that the Nazca plate and the Pacific plate move away from each other as new crust emerges and spreads in both directions. On its Eastern border, the Nazca plate dives under the South American plate.

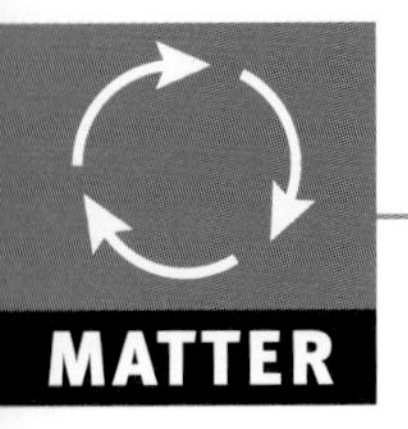

the rock cycle

For our purposes in understanding planet Earth, we need to realize that these plates and their movements explain much more than continents coming together and being pulled apart. The movements of the plates are an important part of the rock cycle.

BIG IDEA

Earth has dry land because the processes that build mountains balance the erosion processes.

Rocks on the Earth's surface are continually broken down by the forces of flowing water, chemical reactions, blowing wind and crushing ice. This broken rock eventually washes into the ocean as sediment. The net effect of this erosion is to lower the surface of the continents to sea level. From the point of view of geological time, mountains crumble rapidly. In the course of just 18 million years, the continents would be reduced to sea level and the oceans would cover the Earth.

Why do we still have continents and mountains that reach miles into the air? Since the continents have existed for hundreds of millions of years, this erosion process must be balanced by a mountain building process. The movements of the plates explain many details of this mountain building.

Sometimes the mountains arise when continental masses collide, as in the case of the Himalayas. Volcanoes demonstrate that mountains are also built from molten material from the Earth's interior. Lava does not just erupt on land. The oceans have submerged mountain ranges that are among the most geologically active regions on our planet. These are places where melted rock constantly flows from the interior to become new crust.

Earth's surface crust is constantly eroded into the oceans, sucked deeper into the Earth's interior and eventually rebuilt into rocks that return to the surface. The same rock material keeps getting used over and over again. When we explore the system of matter on planet Earth, the rock cycle is one reason that we walk around muttering "matter cycles, matter cycles, matter cycles."

ROCK CYCLE

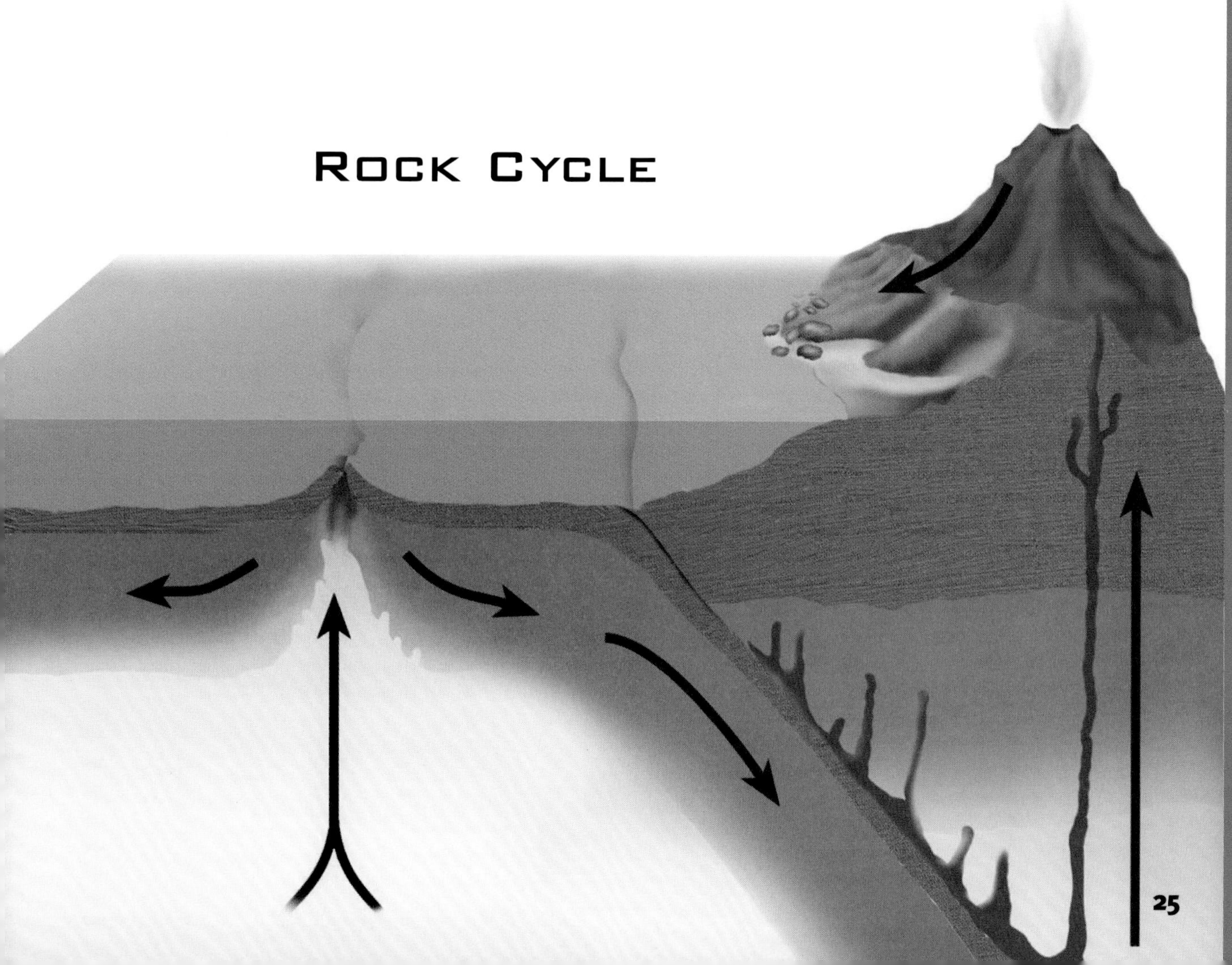

earth's liquid stuff

Water blesses our planet and makes it appear beautifully blue from space. The presence of liquid water clearly distinguishes Earth from all the planets and moons of the solar system. In fact, water covers almost three times as much of Earth's surface as land does.

DID YOU KNOW?

Water covers almost three times as much of the Earth's surface as land does.

Water plays such an important role in our planet that Earth systems scientists extensively study the hydrosphere, the SYSTEM of all Earth's water. This hydrosphere itself can be studied in terms of its subsystem parts – the oceans, frozen water in glaciers and polar ice caps, groundwater, surface fresh water and water vapor in the atmosphere.

The parts of Earth's water system can also be identified as "water reservoirs," places where water occurs. (Scientists use the term reservoir to describe the different places where any substance, not just water, occurs). The water reservoir that holds the most water, 97.25% of all water on Earth, is the ocean reservoir. Check the Water Reservoir table to compare the amounts in other reservoirs, such as glaciers, groundwater, atmosphere and living organisms.

EARTH'S WATER RESERVOIRS		
RESERVOIR	% OF TOTAL	VOLUME IN CUBIC KILOMETERS (KM³)
Ocean	97.25%	1,370,000,000
Ice Caps & Glaciers	2.05%	29,000,000
Groundwater	0.68%	9,500,000
Lakes	0.01%	125,000
Soils	0.005%	65,000
Atmosphere	0.001%	13,000
Rivers	0.0001%	1,700
Biosphere	0.00004%	600
TOTAL	100%	1,408,700,000

We can also compare the different water reservoirs by representing all of the Earth's water as 1,000 milliliters (1 liter) in a beaker. The oceans would contribute most of the 1,000 milliliters. In this comparison, for example, lakes and rivers would contribute just about one drop, and the atmosphere would be a very small part of a drop.

Of course, Earth has much more than 1,000 milliliters. The biosphere, the smallest reservoir in the Water Reservoirs table, contains 600 cubic kilometers.* How much water is that? Enough to fill 60,000 domed stadiums. This means the water in all Earth's plants and animals would fill 60,000 domed football stadiums. So, we're talking about a lot of water even in the smallest reservoir.

0.01 atmosphere
0.1 lakes & rivers
6.8 groundwater
20.5 ice
972.5 oceans

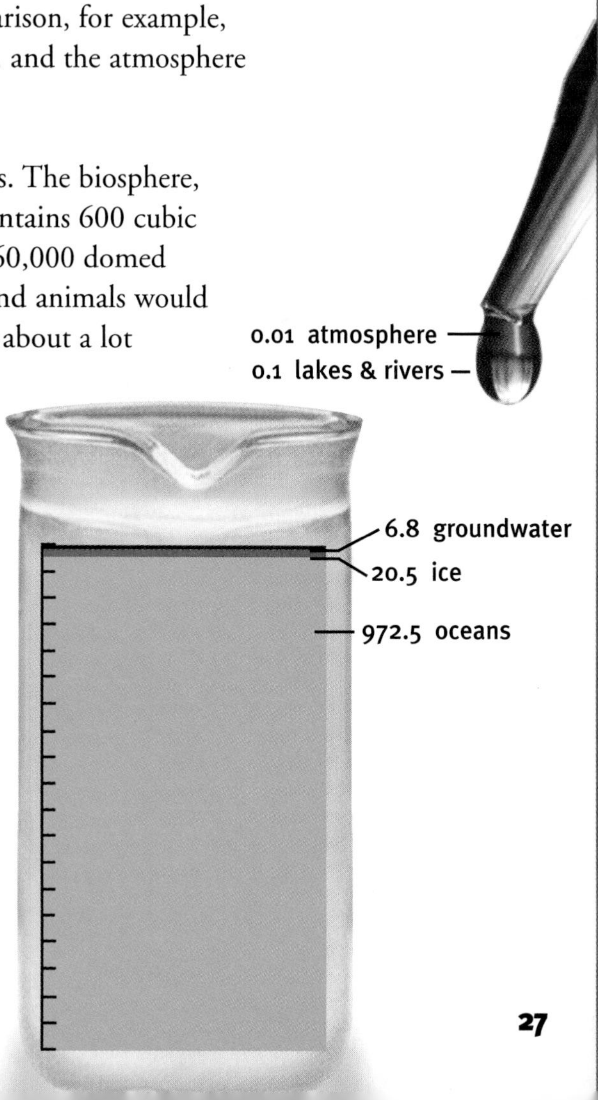

* One cubic kilometer of water fills a cube one kilometer high, one kilometer wide, and one kilometer deep. One cubic kilometer equals about 260,000,000,000 gallons, enough water to fill more than 100 domed stadiums.

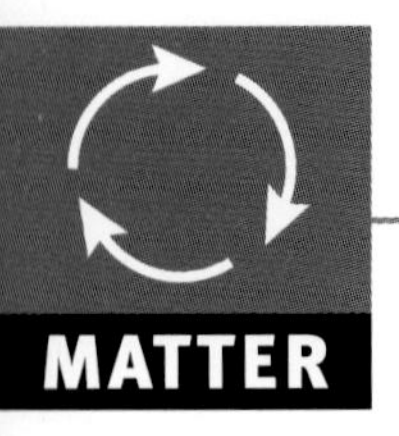

the water cycle

Meet H_2O, a water molecule made up of two hydrogen atoms connected to one oxygen atom. The smallest piece of water is one molecule of H_2O. About one hundred million water molecules placed side by side would stretch all of one centimeter.

Imagine that you are a water molecule here on planet Earth. As we have seen, most of Earth's water exists as a liquid, so you are probably part of an ocean. You have more neighbors than you could possibly count. Even one drop contains an enormous number of water molecules (about 3,000,000,000,000,000,000,000).

"meet a water molecule..."

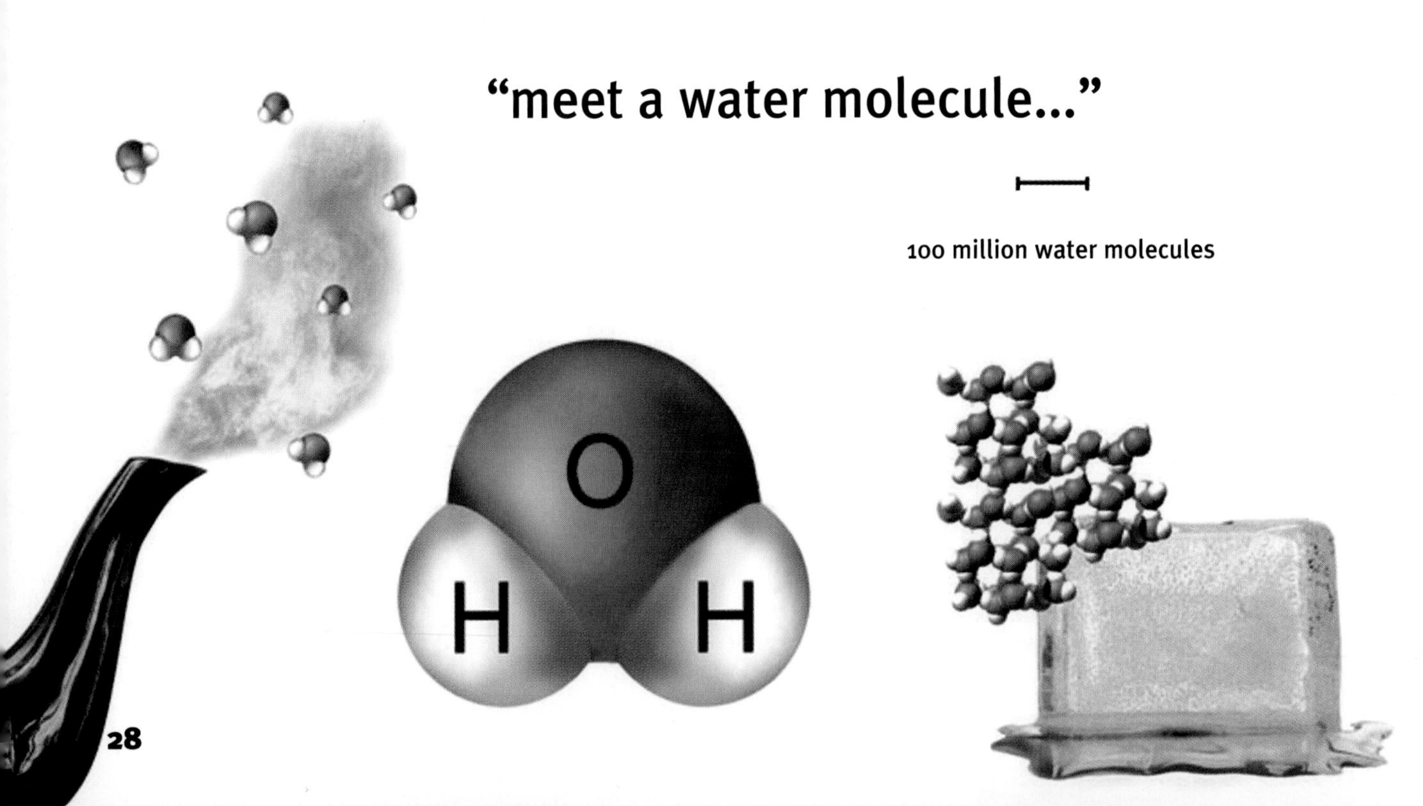

You are constantly moving at speeds on the order of 50 miles per hour, but you don't get anywhere. You are so tightly packed with other water molecules that you constantly crash into them and bounce away. In any given second you travel distances thousands of times your size but it's an endless zigzag of bouncing up and down and back and forth that leaves you very close to where you started. If you had wanted to get anywhere (which, being a water molecule, you don't), you would be very frustrated (good thing you are a water molecule with no desires).

Suddenly a bolt of energy that has traveled 93,000,000 miles from the sun crashes into you. Now you find yourself traveling faster and over much greater distances. And your neighbors have changed. Whereas before all your neighbors were other water molecules, now most of the molecules you bump into are nitrogen or oxygen molecules. Occasionally you smash into other water molecules and rebound away.

Finally you realize what has happened. You have evaporated. What a gas! In fact, you have entered the gas state. You absorbed enough solar energy to escape from the close attraction of all those other water molecules back in your old liquid neighborhood. Amazingly brave water molecule, you jumped out of the liquid state into the vast unknown, into thin air.

Here on planet Earth, water molecules behave like this all the time. They evaporate from the liquid state into the gas state. They also perform the opposite trick. When a lot of water molecules zooming around in the gas state attract each other, they can stay connected as a very, very tiny drop of liquid water. They connect with other very tiny drops and before they know it, gravity forces them to fall to the ground in the liquid state. They have precipitated as rain or snow.

Returning now to thinking about water reservoirs on planet Earth, we see that this evaporation and precipitation causes water to move from one reservoir to another. A water molecule does not tend to stay trapped in the same old reservoir. Over the course of time, it changes both its physical state (gas, solid, liquid) and its physical location (ocean, atmosphere, glacier, river).

Let's examine a reservoir in the Water Cycle illustration (page 31). For the ocean, 434 units (each unit is equal to a thousand cubic kilometers of water) leave per year via evaporation. However, 398 of those units return directly to the ocean as precipitation (rain on the ocean). The remaining 36 units get blown over to the land where they are deposited primarily as rain and snow. If they did not return to the ocean, then the ocean would progressively lose water. However, that is not what happens. Over the course of a year, 36 units of water flow as runoff from the land into the ocean. The net result is that just as much water enters the ocean as leaves it and the total volume of the ocean does not change. The amount of water in the atmosphere also remains constant because the amount entering equals the amount leaving.

From a long term global perspective, we see that the same water molecules are used over and over again. The hydrosphere, planet Earth's water system, is a closed system. No new water enters the hydrosphere. No used up water leaves the hydrosphere. The same water keeps moving from one reservoir to another, going round and round, leading to the name we give this phenomenon – the Water Cycle. When we explore the system of matter on planet Earth, the water cycle is another reason that we walk around muttering "matter cycles, matter cycles, matter cycles."

HOW LONG A WATER MOLECULE STAYS IN THE OCEAN COMPARED TO HOW LONG IT STAYS IN THE ATMOSPHERE			
RESERVOIR	AMOUNT OF WATER IN RESERVOIR	HOW MUCH ENTERS AND LEAVES PER YEAR	HOW LONG AN AVERAGE WATER MOLECULE STAYS IN THE RESERVOIR
Ocean	1,370,000,000 cubic kilometers	434,000 cubic kilometers	1,370,000,000 divided by 434,000 **equals 3,160 years**
Atmosphere	13,000 cubic kilometers	505,000 cubic kilometers	13,000 divided by 505,000 equals 0.026 years equals **about 9 days**

As we have seen, the different reservoirs of the water cycle can differ greatly in the amount of water they contain. They also differ in the rate at which it enters and leaves. The chart above shows that a water molecule stays in the

ocean about 3,000 years, while it stays in the atmosphere only 9 days. The same water cycles over and over through the various reservoirs.

To sum up, Earth's liquid stuff, its hydrosphere, exists in reservoirs that are connected through the water cycle. These reservoirs differ greatly in their size and the rate at which material enters and leaves them.

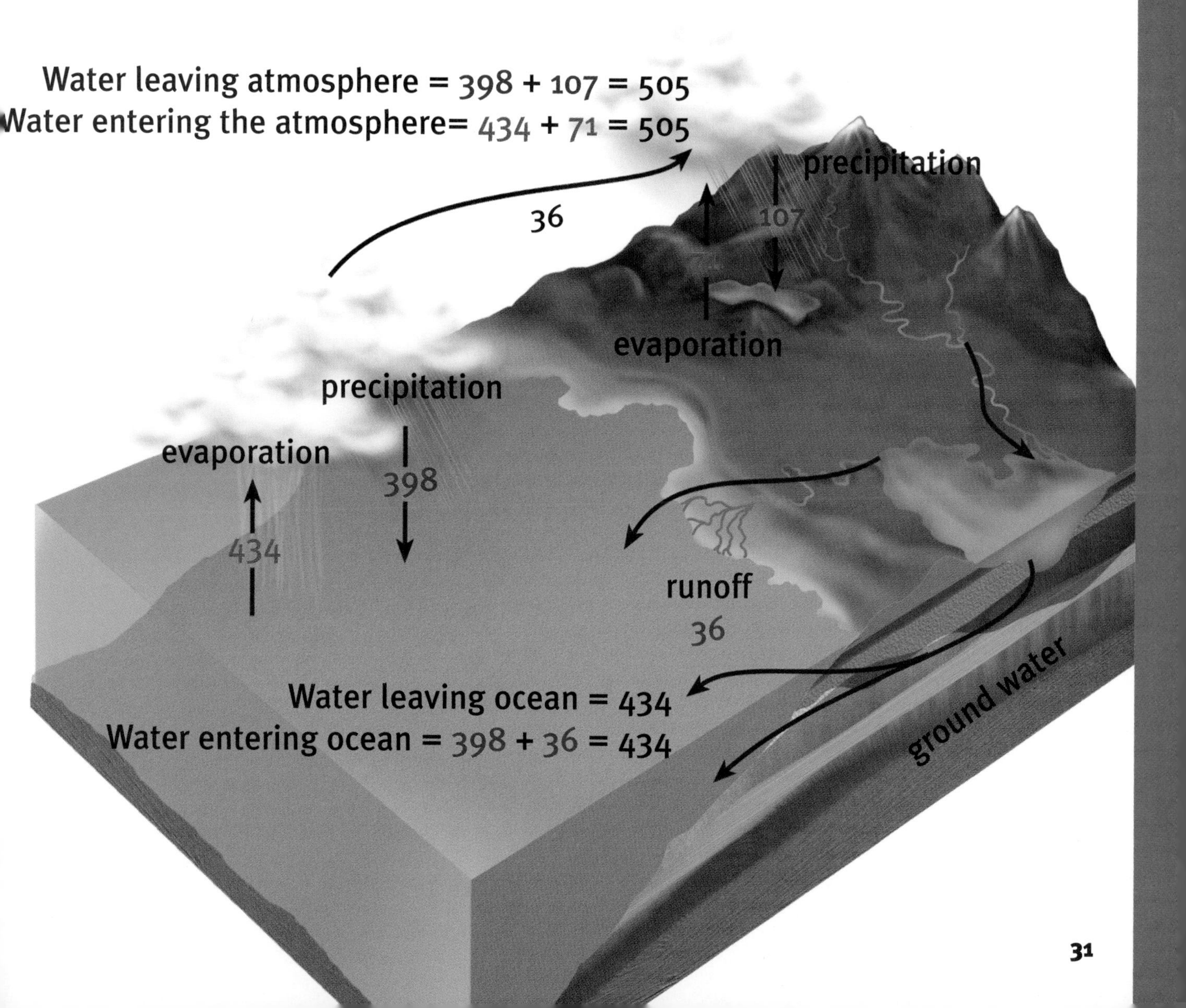

Here is another way to understand the water cycle. Think about one of our ancestors who lived in Africa a million years ago. Or think about a dinosaur that lived 70 million years ago. Or consider a buffalo that roamed the American Midwest millions of years before the arrival of humans. No matter which you choose to bring to mind, that organism drank water throughout its life. This water was present in every drink and in every grain, fish or flesh that was consumed. The water molecules became part of that organism's body and then flowed back into the world as blood, sweat, urine and exhaled water vapor.

Now, fill a glass with water. This glass that you hold in your hand today has more than ten million water molecules that once passed through the body of the buffalo, more than ten million water molecules that passed through the dinosaur and more than ten million water molecules that passed through one of our African ancestors! The water that we drink connects us intimately with the living beings that inhabited the planet before us, that inhabit Earth today and that will inhabit it in the future.

BIG IDEA

The water that we drink connects us intimately with the living beings that inhabited the planet before us, that inhabit Earth today and that will inhabit it in the future.

earth's gas stuff

Earth's atmosphere is a very thin layer of air that protects and sustains us. At the top of tall mountains, most of us experience problems breathing due to the thinning of Earth's atmosphere. The higher we go, the fewer the gas atoms in the atmosphere and the more it resembles the emptiness of outer space.

BIG IDEA

The atmosphere is the most sensitive and changeable of Earth's spheres.

Compared to the geosphere and the hydrosphere, the atmosphere is the most sensitive and changeable of Earth's "spheres." It can change quickly because it is comparatively very small. In terms of mass, the whole Earth system contains a million times more solid stuff than gas. Therefore, if a small part of Earth's solid stuff changes to gas and enters the air, it can have a major effect on the atmosphere.

Nitrogen accounts for almost four fifths (78%) of the gas in the atmosphere. Oxygen at 21% accounts for almost all the rest. The rest of the gases in the atmosphere are present in much smaller amounts, with the most important being carbon dioxide at about 0.03%. As we all experience, the atmosphere also has varying amounts of water vapor, depending on the location and the weather at any particular time. The same volume of warm air above a tropical rainforest can contain a hundred times more water than the cold, dry air over the Antarctic.

I hope you will not be surprised that the nitrogen, oxygen and carbon located in Earth's atmosphere each participate in a matter cycle. By now you should expect that matter on Earth is used over and over again. Everything on the planet is made of atoms and these atoms on Earth are neither created nor destroyed. The same atoms keep combining, separating and recombining with each other.

We will focus here on one of Earth's most important cycles, the carbon cycle. Since all Earth's organisms are carbon-based life forms, we should pay careful attention to the carbon cycle. Plants and animals actively participate in this cycle by exchanging carbon dioxide out of and into the atmosphere. Currently humans add 7 billion extra tons of carbon per year into the atmosphere by burning fossil fuels and forests.

the carbon cycle

The carbon cycle is harder to understand than the water cycle. With the water cycle, we are talking about the same molecule (H_2O). In going through the water cycle, these H_2O molecules change in their physical location and in their physical state (gas, liquid and solid). In the carbon cycle, the carbon atoms change not only in their physical condition but also in their chemistry.

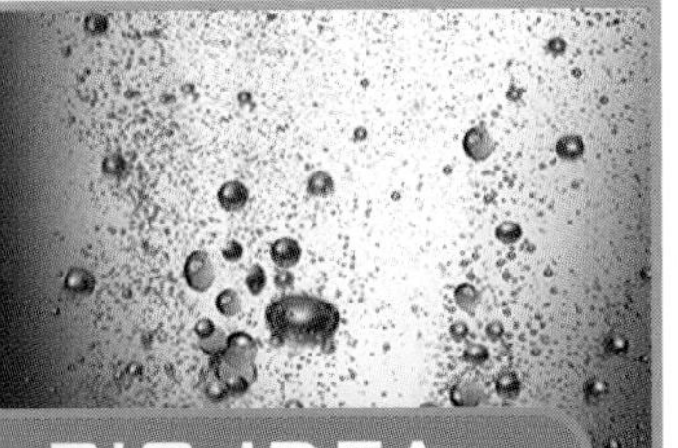

BIG IDEA

In the carbon cycle, carbon atoms change their chemical partners, physical location and physical state.

Carbon in the atmosphere is mostly present as the gas carbon dioxide (CO_2; one carbon combined with two oxygens). In living and decaying matter, carbon is present as carbohydrates and proteins where it bonds with oxygen, hydrogen and other elements in a huge number of different chemicals. In the ocean, it is present mostly as bicarbonate salts (bicarbonate is a combination of carbon, oxygen and hydrogen that is also commonly found on supermarket and kitchen shelves in the form of baking soda). With the carbon cycle, we see the same carbon atoms changing their chemical partners as well as their physical location and physical state (gas, liquid and solid) as they flow from one reservoir to another.

The carbon cycle illustration on the next page and chart below show five major carbon reservoirs on Earth. Each of these reservoirs is an important location where carbon exists on our planet. They are the Atmosphere, Biomass, Ocean, Sedimentary Rocks and Fossil Fuels. The numbers next to the arrows represent the rate (in billions of tons per year) at which carbon enters and leaves each of those reservoirs. There is some uncertainty in the exact value of these numbers but the relative amounts are correct.

CARBON RESERVOIRS AND THE ATMOSPHERE
(1 gigaton = 1 billion tons)

RESERVOIR	FORM OF CARBON	AMOUNT OF CARBON	FLOW RATE WITH ATMOSPHERE	HUMAN EFFECTS ON ATMOSPHERE
ATMOSPHERE	Carbon Dioxide (gas)	760 gigatons	Not applicable	Greenhouse gases are increasing
BIOMASS (mostly carbon in plants and soil)	Sugar, Cellulose, Protein, etc. (solid, dissolved)	2,000 gigatons	Per year, about 110 gigatons flow in each direction	Burning forests release about 1 gigaton per year
SEDIMENTARY ROCK	Carbonates (solid)	50,000,000 gigatons	About 0.05 gigatons per year	Negligible
OCEAN	Mostly dissolved bicarbonate salts	39,000 gigatons	About 90 gigatons per year; we think the ocean is absorbing more than it releases.	Negligible
FOSSIL FUELS	Methane (gas), Petroleum hydrocarbons (liquid), Coal (solid)	5,000 gigatons	Natural background rate	About 6 gigatons per year above natural background rate through burning methane, oil and coal

It is easiest to understand the carbon cycle by exploring how each of the different reservoirs interacts with the atmosphere. We will examine how the atmosphere interacts with life, rocks, oceans and fossil fuels. As we shall see in the next chapter, atmospheric carbon dioxide plays an important role in determining Earth's climate. This crucial role provides another reason to focus on the atmosphere when we explore the carbon cycle. The atmosphere contains 760 billion tons of carbon (as measured in the late 1990s), almost all of it present as carbon dioxide. This CO_2 currently makes up 0.035% of the atmosphere, a small percentage but essential for life as we know it.

The most well-known part of the carbon cycle involves life on the continents (Land Biomass). When we think about life, we usually focus

Carbon cycle

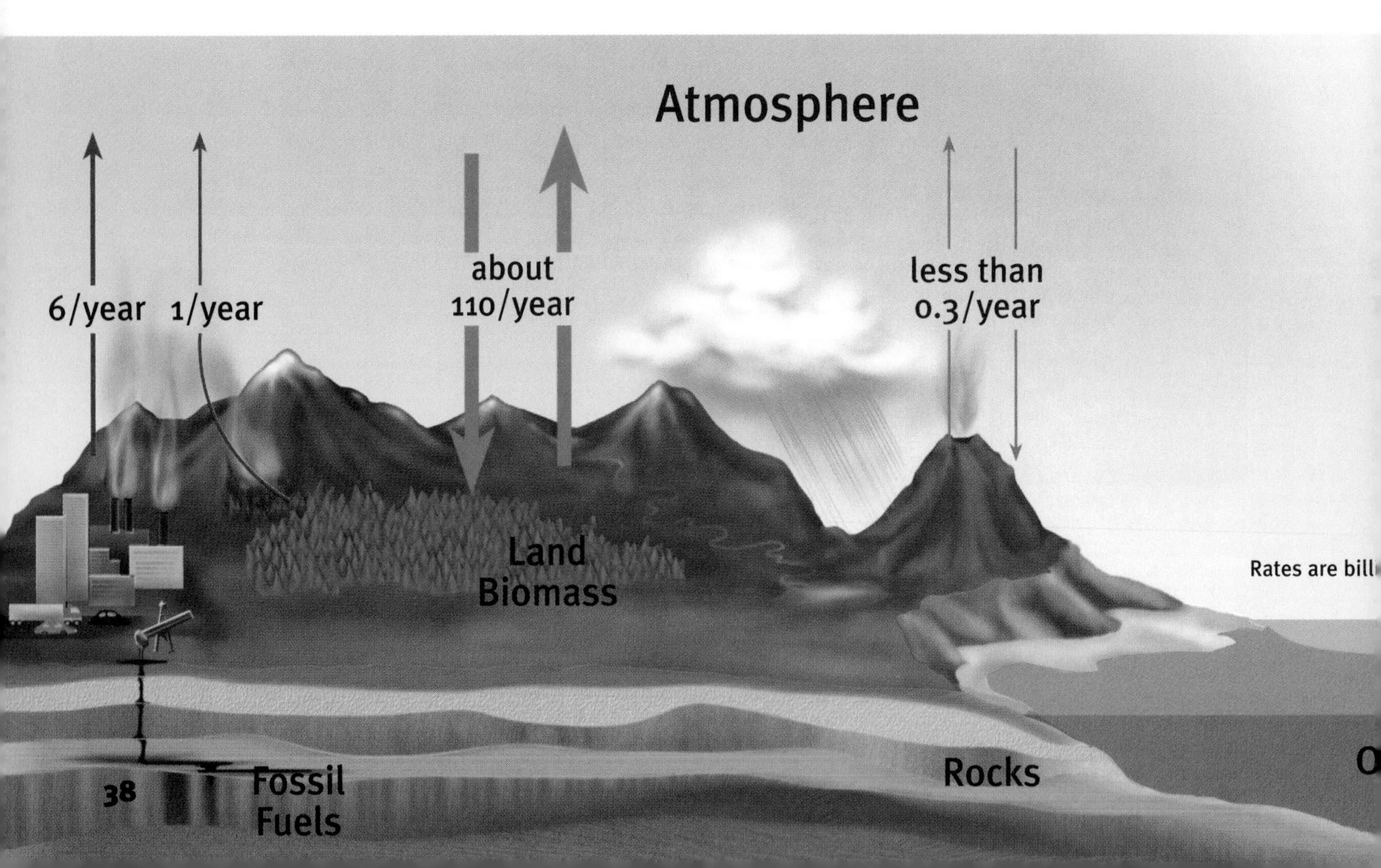

on animals. However, this part of the carbon cycle is really about plants and trees and how they change carbon dioxide gas into living matter. Eventually this living matter is used by plants, animals and decomposers to return the carbon to the atmosphere, again in the form of carbon dioxide gas (described in more detail in Chapter 4).

During an average seven year period, all the carbon in the atmosphere will leave it and become part of land-based living organisms. Over the same time period, an equivalent amount of carbon in these living or decaying organisms will return to the atmosphere as carbon dioxide. The net result from these interactions is that the amount of carbon in the atmosphere remains constant even though carbon atoms constantly leave and enter the atmosphere as they cycle through living organisms.

The oceans are a very significant reservoir in the carbon cycle, containing about 20 times more carbon than land biomass and 50 times more carbon than the atmosphere. This ocean carbon is present mostly as dissolved bicarbonate salt. The annual rate at which atmospheric carbon enters and leaves the ocean is similar to the rate of exchange with land biomass. In other words, about every seven years all the carbon in the atmosphere will leave it and become part of the ocean. Similarly, about every seven years approximately the same amount will leave the ocean and return to the atmosphere.

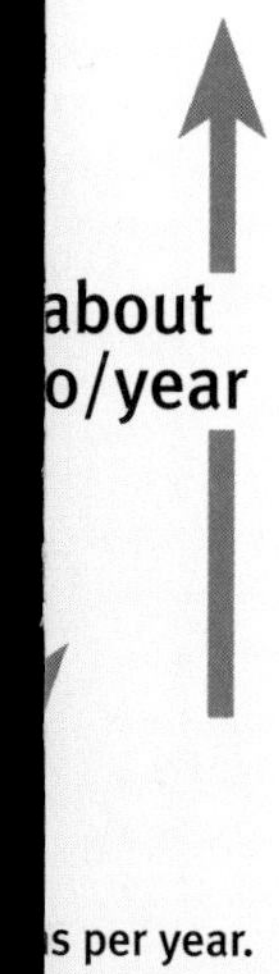

Rocks contain the vast majority of Earth's surface carbon, more than 50,000 times as much as the atmosphere. However, this huge store of carbon interacts with the atmosphere at a much slower rate. In one direction, a process called weathering removes carbon from the atmosphere. In the other direction, hot springs, volcanoes and other upheavals return carbon to the atmosphere from Earth's interior. We met this part of the carbon cycle before when we explored our old friend the rock cycle.

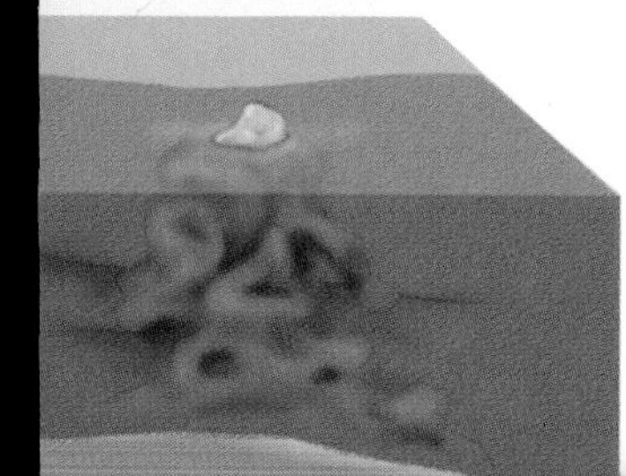

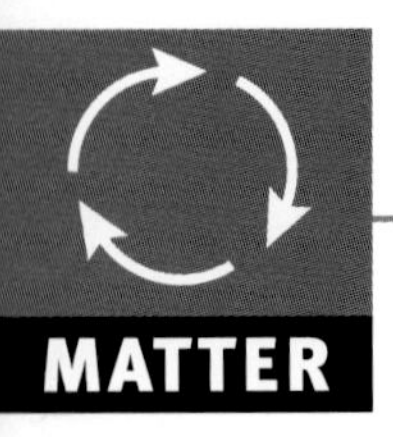

today's carbon cycle

When we examined the water cycle, we noted how the amount of water in each reservoir currently remains constant. The same amount of water evaporates from the ocean as eventually returns to it in the form of rain or runoff from the land. The same amount of water enters the atmosphere through evaporation as leaves it through precipitation.

Cycles, such as the water and carbon cycle, have changed during Earth's history. For example, during an ice age, a large amount of water leaves the ocean and remains on land in the form of glaciers. As a result, the oceans dramatically decrease in size. Islands may become part of the mainland and land bridges may connect previously separated continents.

Today's carbon cycle is changing, but this time as the result of human actions. The carbon cycle illustration on the previous pages shows that humans are adding extra carbon to the atmosphere by burning forests and fossil fuels.

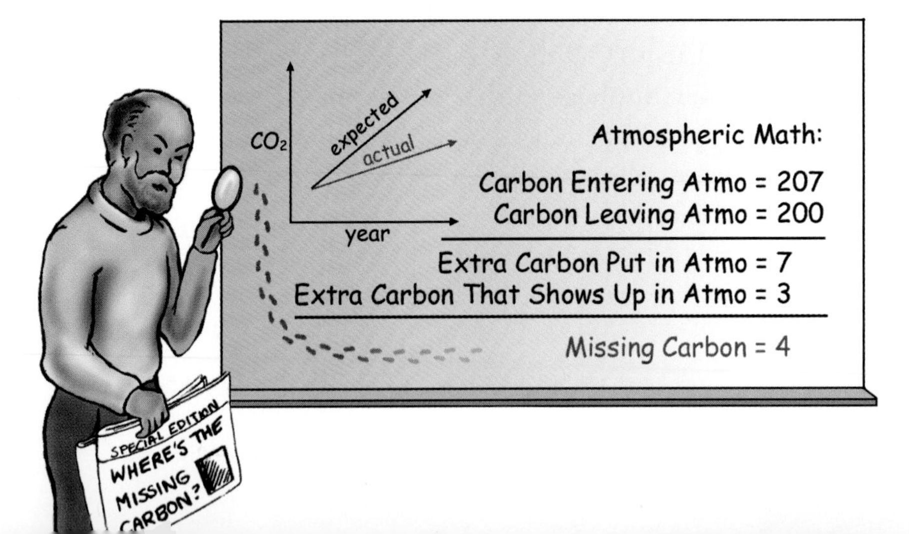

The burning of fossil fuels (oil, coal and natural gas) accounts for the greatest amount of carbon that humans are adding to the atmosphere. Approximately 6 billion tons of carbon in the form of carbon dioxide enters the atmosphere due to the burning of fossil fuels for transportation, heating, cooking, electricity and manufacturing. The carbon in fossil fuels comes from living organisms that were buried millions of years ago. This fossil fuel carbon had been stored deep underground in oil, coal and gas deposits, and had therefore been locked out of the current atmospheric carbon cycle.

The global carbon cycle is currently not in balance. By burning forests (1 billion tons) and fossil fuels (6 billion tons), humans add more than 7 billion tons of carbon per year to the atmosphere. What happens to all that carbon? Currently about 3 billion tons remain in the atmosphere with the result that the concentration of carbon in the atmosphere has increased 25% in the last century.

What happens to the rest? Since Earth is essentially a closed system for matter, the extra carbon added to the atmosphere has to end up somewhere. As you can imagine, it is difficult to measure all the carbon in the ocean, trees or rocks. While nobody can prove the location of this Missing Carbon, scientists have some evidence that the oceans are absorbing about half of this extra carbon and that growing forests may be absorbing the other half.

At current rates of burning fossil fuels, the carbon in the atmosphere may double in amount sometime around the year 2050. If the ocean and growing forests do not continue to absorb more than half of the extra carbon, this amount could increase even faster. The next chapter gives us a very strong reason for caring whether and how fast atmospheric carbon increases in amount.

Explore Chapter 2 on the web...

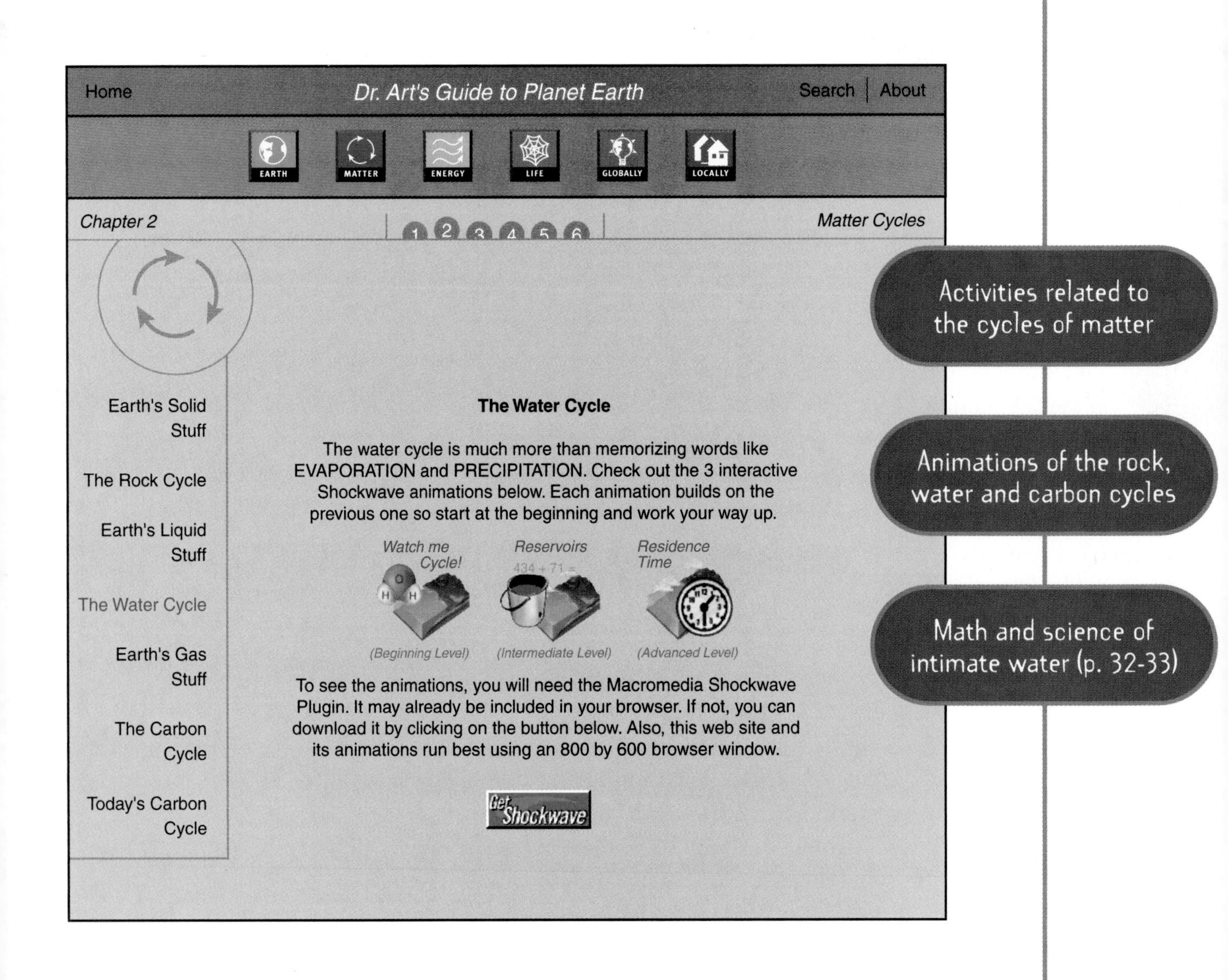

- Activities related to the cycles of matter
- Animations of the rock, water and carbon cycles
- Math and science of intimate water (p. 32-33)

www.planetguide.net

energy flows

Earth's Energy

Part of a Larger System

Energy from the Sun

The Greenhouse Effect

Earth's Internal Energy

Earth's Energy Budget

earth's energy

As we begin to explore Earth's energy, we confront a question that we took for granted in the first chapter. What does this word "energy" mean? What is energy?

BIG IDEA

Energy is neither created nor destroyed.

You might be surprised to learn that scientists don't really know what energy is. We can measure it very precisely. We can accurately predict how it will change from one form to another. Still, if you question a physicist and insist on getting a definition of energy, the answer you hear will sound more like a riddle than a definition:

"Whenever something happens, there is a property of the system that does not change in amount. We give the name energy to that property of a system that never changes in amount."

In science books they usually call this the Law of Conservation of Energy. The most user-friendly way to say it is:

"Energy always remains constant in amount. Energy is neither created nor destroyed."

At first sight, this scientific law does not fit our experience of the world. We fill the car's gas tank on Monday and 350 driving miles later we return on Friday to refuel. Our local energy company charges us for the oil or natural gas that we use to heat our home. If we refuse to pay the bill and send a letter to the company arguing that scientific law tells us that we did not use up the energy, what do you think the answer will be?

The Law of Energy Conservation takes a much broader view of energy than we normally do. When we heat our home, we pay attention only to the fuel and the heat in the house. The Energy Conservation Law follows the heat after it leaves the house, watches it escape through the atmosphere, spread into outer space, and notices that the heat continues to exist forever - it is never destroyed. Further, the amount of heat energy exactly equals the amount of chemical energy released from the fuel (such as gas, oil or wood). The company does not bill us because we destroy energy. We pay the electric and gas bill because we use a particularly convenient form of stored energy and change the energy into a form that is much less useful.

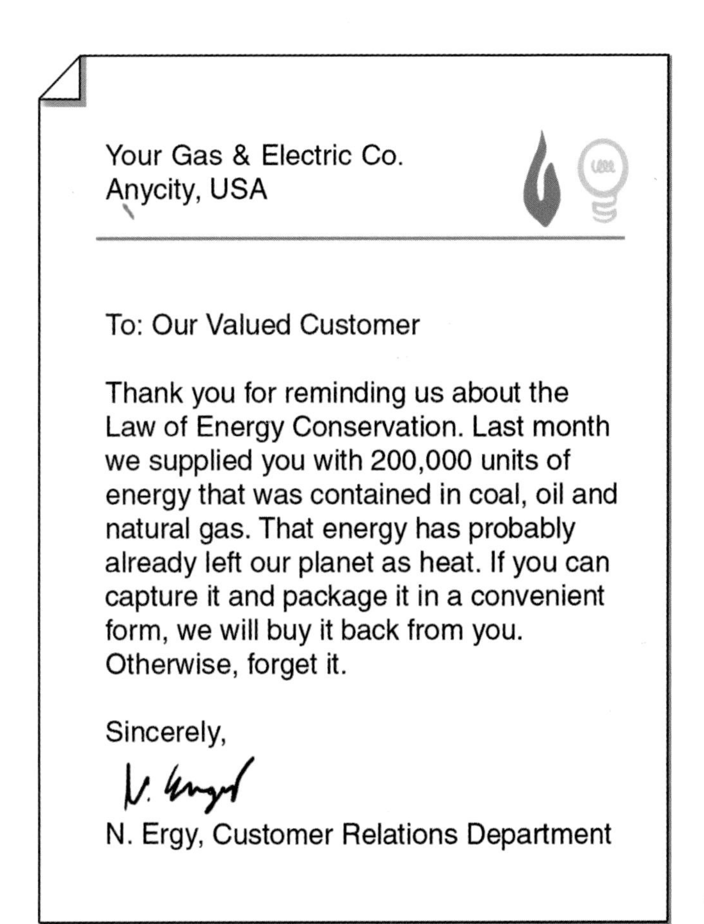

Your Gas & Electric Co.
Anycity, USA

To: Our Valued Customer

Thank you for reminding us about the Law of Energy Conservation. Last month we supplied you with 200,000 units of energy that was contained in coal, oil and natural gas. That energy has probably already left our planet as heat. If you can capture it and package it in a convenient form, we will buy it back from you. Otherwise, forget it.

Sincerely,

N. Ergy, Customer Relations Department

part of a larger system

Back in Chapter 1, we introduced systems thinking as a way to understand any system, especially a complicated one like planet Earth. We said that three systems questions often help us analyze the system that we are exploring. When we looked at Earth's matter, we mostly used the first systems question: "What are the parts of the system?"

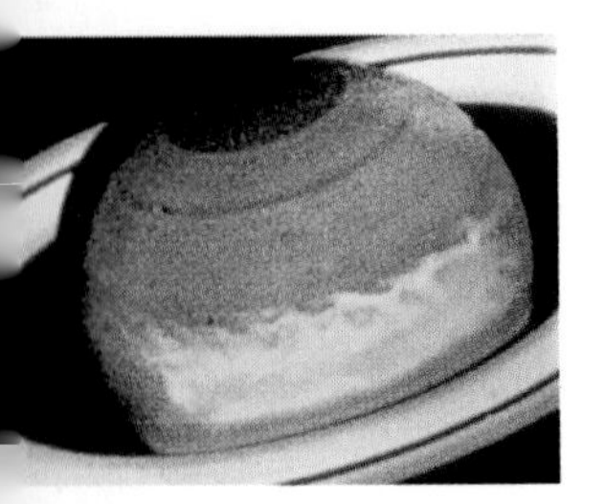

We examined Earth's solid, liquid and gas parts and discovered that they all participated in cycles. We concluded that matter on planet Earth cycles, that Earth is essentially a closed system for matter.

What if we asked the same system question about Earth's energy? Looking for Earth's energy "parts," we might run around measuring the wind, hot springs, volcanoes, waterfalls and humans making fires. Then if we take a break and relax on the beach, we would realize that we had ignored Earth's most important source of energy.

It's way out there. It's not one of Earth's parts. It provides our planet with 15,000 times more energy than all our societies consume. Of course, it's the sun. To understand energy in the Earth system, we need to focus on a different systems question. Instead of looking at the parts of the Earth system, we need to ask the third system question: how is Earth itself part of larger systems?

And the answer is as simple as the question – Earth is part of the solar system. The sun provides virtually all the energy to keep our planet warm and sustain life.

Since our solar system has many other planets, we also discover how important it is to be close to the sun but not too close. When the planets first formed, the areas closer to the sun were too hot for anything but rocky materials to solidify. So these inner planets (Mercury, Venus, Earth and Mars) are mostly rock. In contrast, the outer planets (Saturn, Jupiter, Uranus and Neptune) were cool enough to keep materials such as methane and ammonia and they became very large, consisting mostly of atmospheres containing these and other gases.

Some people have called the third planet from the sun the Goldilocks planet. In the children's story of Goldilocks and the three bears, she found the chair that was not too big or not too small and she ate the porridge that was not too hot or not too cold. Earth is not too close to the sun, not too far, not too hot, and not too cold. Earth is just right.

energy from the sun

Since energy never changes in amount, you might think that it does not live up to its name, that it is pretty dull. Well, you would be wrong. Energy is, well, energetic. While it does not change in amount, it changes forms very readily.

Check out what happens to the solar energy that reaches Earth. About 30% is immediately reflected back as light to outer space. Most of this light bounces off the clouds and never reaches the surface. Some of it reaches the

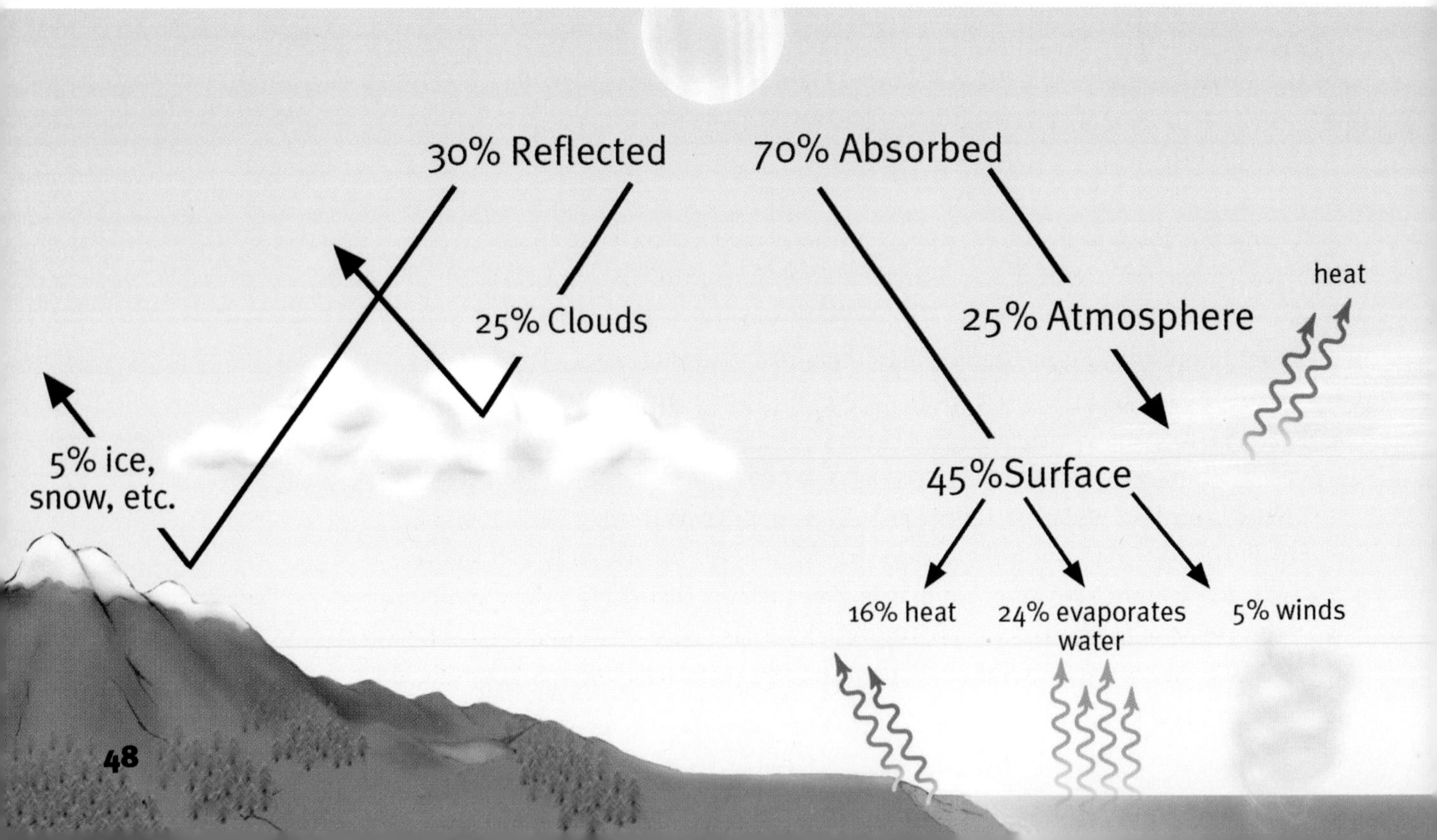

surface but bounces off snow and water, also leaving the Earth system in the form of light. This reflected light makes Earth visible from space.

The remaining 70% of the sunlight that reaches Earth is absorbed. As shown in the illustration, this absorption occurs in a number of ways. Most of it is absorbed by solid materials and water and is immediately converted to heat. We all experience this phenomenon when sunlight warms our bodies. What we don't usually consciously experience is that this heat energy radiates from our bodies. Any material that is heated by the sun will then radiate heat outwards. Sometimes we see this heat as in the shimmering waves of air above a hot pavement. Eventually that heat radiates through the atmosphere and leaves planet Earth by flowing to outer space.

A large amount of the solar energy evaporates water, thereby powering the water cycle. Water absorbs this energy as it changes from the liquid state to the gas state. Water vapor then leaves the oceans and enters the atmosphere. However, when this water vapor condenses back to the liquid state (rain), the same amount of energy is released in this condensation process as the amount that was absorbed in the evaporation. This energy is now released as heat that escapes to the atmosphere and eventually to outer space. So even the incoming sunlight that powers the water cycle also eventually leaves Earth in the form of heat.

The same fate awaits the solar inflow that is initially converted to the moving energy of wind, waves and currents. The same fate awaits the tiny but crucial amount (0.08%) that plants capture and convert to chemical energy. The wind rubs against a cliff and some of its energy changes to heat. A cow eats grass and converts the plant chemical energy to body heat which radiates to outer space.

No matter how the energy is absorbed or changes its form, it never increases or decreases in amount. That is one key characteristic of energy - it is never created or destroyed. Another key characteristic is that energy changes form, eventually changing to heat energy. All the solar energy that is absorbed on Earth in one form or another eventually changes to heat energy that radiates to outer space.

the greenhouse effect

I have used the word "radiate" as if everyone knows what it means. In the last section, you might even have become annoyed with me repeating that "heat energy radiates to outer space." Whoops, I just said it again. What do we mean by that phrase?

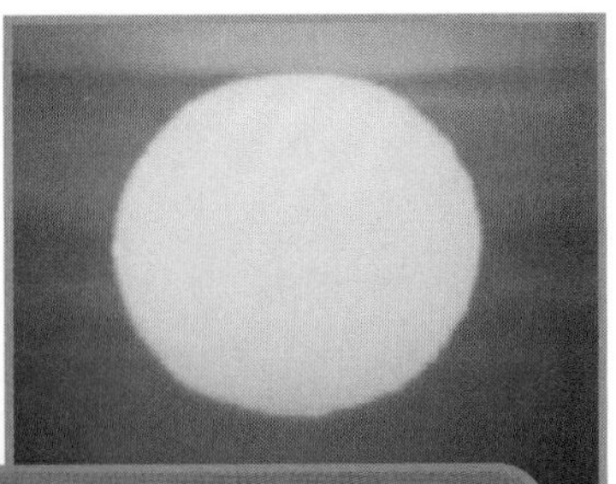

BIG IDEA

We are bathed in electromagnetic energy. It bounces off, is absorbed by and flows through us.

You may also have noticed that I have avoided using scary sounding scientific words in this book. Well, don't freak out, but to explain solar energy and heat radiating, I need to use one – the electromagnetic spectrum. Many very familiar forms of energy are electromagnetic in nature. Examples include green light, red light, microwaves, radio waves, ultraviolet light and X-rays.

We call them electromagnetic because, guess what, they have electrical and magnetic properties. More importantly, they all travel at the speed of light (in other words, as fast as anything can go), do not lose energy as they travel (even over huge distances, such as from the sun to Earth) and travel like waves.

Some of these forms of energy even have wave or ray right in their name. All of them travel as waves and the feature that makes them different from each other is their wavelength. Each of these different forms of electromagnetic energy possesses its own characteristic wavelength.

The wavelength is what makes green light different from a radio wave and different from an x-ray. The wavelength of x-rays is about a thousand times *shorter* than green light, while the wavelength of radio waves is about a thousand times *longer* than green light. This spectrum (meaning a broad range going from one end to the other) includes electromagnetic waves that differ by more than a billion times in the size of their wavelengths.

Which brings us to our sun. The sun is not boring. It doesn't give off just one wavelength. It radiates energy across a fairly broad range of wavelengths. You know this because you have seen rainbows, natural examples where some of the sun's light separates into its different wavelengths. The longer waves (which we see as red) appear at the top, and the shorter waves (blue) at the bottom.

The sun radiates about half of its energy in the visible part of the electromagnetic spectrum. We have evolved so we can see half of the sun's radiant energy ranging from short wavelengths that we see as violet to wavelengths about twice as long that we see as red. The sun emits 40% of its energy in the infrared (IR) region (longer than red wavelengths, which some animals such as rattlesnakes can see). It also emits about 10% of its radiation as ultraviolet (UV) rays (shorter than violet, which some animals such as bees can see).

A radio wave can have a wavelength a billion times longer than an X-ray.

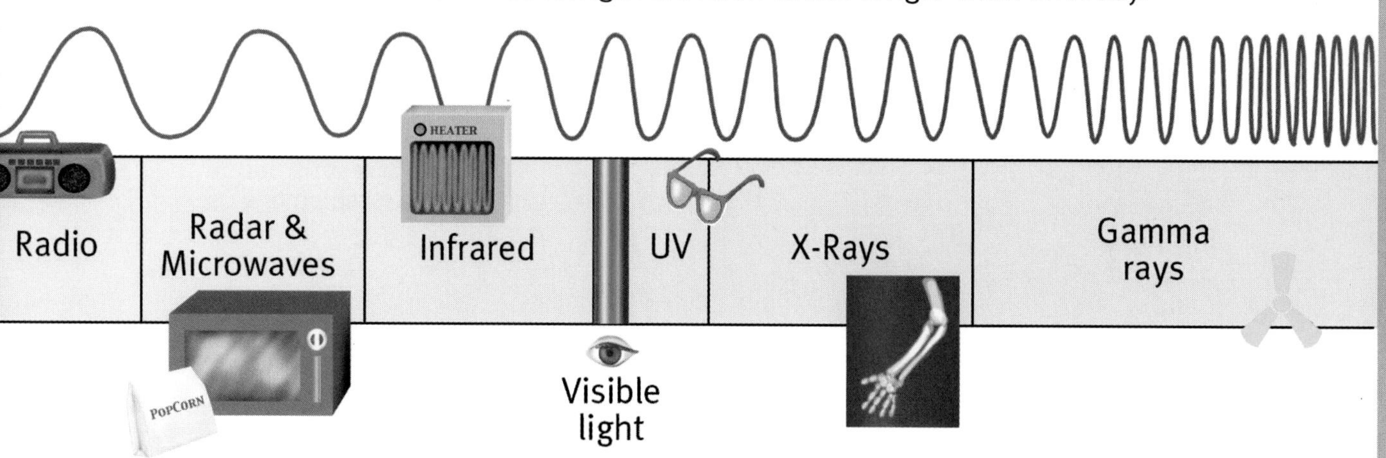

Now that we have met and become comfortable with the electromagnetic spectrum, we can discuss and understand Earth's famous greenhouse effect. Here we focus on the visible sunlight and what happens to it. The visible light rays that do not bounce off clouds pass right through the atmosphere. Most of these rays then crash into water, rocks, soil, sand, buildings, roads and organisms.

What happens to a rock when light rays crash into it? The energy from the light makes the rock molecules move faster. In other words, the rock becomes warmer. If something feels warm to us that means its molecules have more energy and are moving faster than when we experience the same object as being colder.

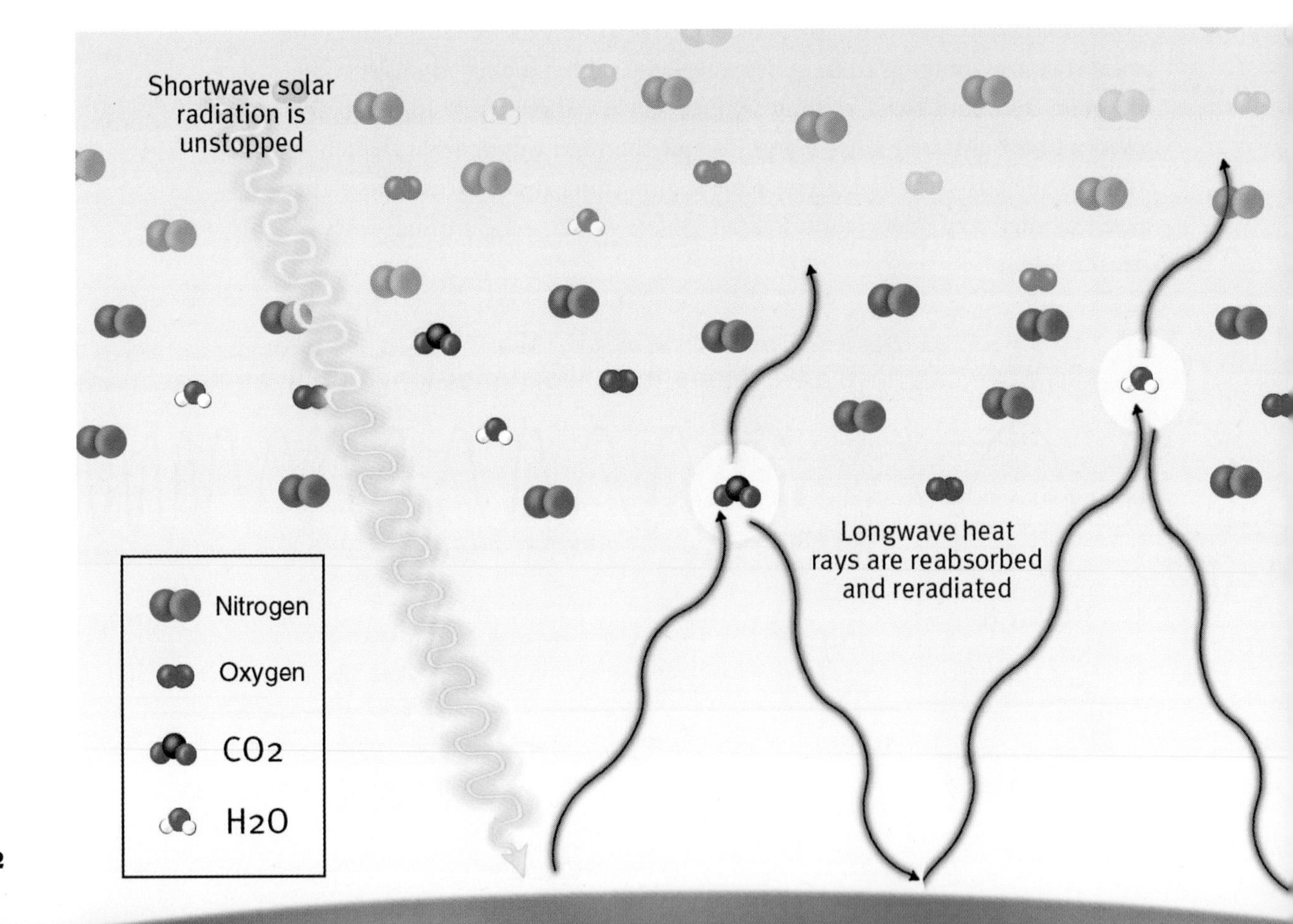

Do warm objects stay warm? No, they tend to get colder. Remember energy does not stay in one place. A hot object such as a rock in the sun gives off some of its energy as electromagnetic rays. These are infrared waves (longer than red) that carry heat away from the rock. When you feel heat from a fire or any other hot object that you are not physically touching, you are generally experiencing infrared heat rays that have radiated to you from the fire or the hot object.

Infrared rays from all over the planet radiate out to the atmosphere. The energy started as visible light from the sun that passed right through the atmosphere, crashed into objects and heated them, and then traveled away from the objects in the form of longer wavelength infrared (IR) radiation.

Unlike the shorter wavelength visible light, this infrared radiation does not just pass through the atmosphere. Certain gases in Earth's atmosphere absorb the radiated heat energy. These atmospheric greenhouse gases (mostly water vapor and carbon dioxide) then radiate that heat energy so that half of it returns to Earth where it is absorbed before being reradiated back out to the atmosphere. The net result is that heat energy stays longer within the Earth system than it would if water vapor and carbon dioxide were absent from the atmosphere.

These atmospheric gases are called greenhouse gases because they let light rays pass through but absorb heat rays. We are very lucky that Earth has this greenhouse effect that slows the rate at which heat leaves the Earth system. As a result, Earth is about 33 degrees Celsius (60 degrees Fahrenheit) warmer than it would be in the absence of the greenhouse effect. Without the greenhouse gases in the atmosphere, Earth's average temperature would be well below the freezing point of water. It would be colder than any of the ice ages that Earth has experienced.

earth's internal energy

Until now we have emphasized that the sun provides virtually all the energy for planet Earth. This solar radiation keeps Earth warm, powers the wind, drives the water cycle and provides the energy for almost all Earth's creatures.

Back in Chapter 2 we explored a different kind of energy and some important roles that it plays in the Earth system. We saw Africa and South America being dragged apart while the Indian subcontinent traveled 4,000 miles and crashed into Asia. What energy source provides the power to move continents? Even the sun, our greatest energy source, does not move continents.

Volcanoes, earthquakes, geysers and hot springs provide clues to the answer. Earth's interior is hot enough to melt rocks and metal. This heat energy constantly moves as it slowly travels to the cooler surface, to the atmosphere and eventually to outer space. The hottest material deep in Earth's interior slowly rises toward the surface and cooler material sinks toward the interior. These patterns of heat flow cause Earth's plates to move (remember the plates?) with their resulting earthquakes, volcanoes and moving continents.

When Earth first formed, the heat was so great that the entire planet consisted of melted rock and metal. Since that time is has been cooling with heat rising to the surface and radiating to outer space. In addition new heat is continually being generated internally since Earth's materials contain radioactive elements that break down and release heat in this decay process. The heat from radioactive decay plus the heat remaining from Earth's formation provide the power to move the plates and drag continents apart or crash them into each other.

How large is this energy source that can move continents, blow the top off Mount St. Helens and create mountains such as Everest? The chart below compares all of Earth's energy flows. If we give the value of 1 to the amount of energy that human societies consume, then earth's internal energy flow is 2.5 times as large and the sun provides 15,000 times that amount.

COMPARING AMOUNTS OF ENERGY	
TYPE OF ENERGY	RELATIVE AMOUNT
Human Societies	1.0
Internal	2.5
Solar	15,000.0

How can something that seems so weak have such large effects? A Chinese saying provides an important clue: "The journey of a thousand miles begins with a single step." The flow of internal energy moves the plates at rates of only centimeters per year. Yet, over hundreds of millions of years, those centimeters add to up to thousands of kilometers.

So, although Earth's internal geothermal energy contributes very little energy compared to the sun, it is a very important part of Earth's energy budget. Tyler Volk, an Earth Systems scientist at New York University, has written that while the energy from the sun through falling rain and blowing wind can make molehills out of mountains, only Earth's internal energy can make mountains out of molehills.

BIG IDEA

Only Earth's internal energy can make mountains out of molehills.

ENERGY

earth's energy budget

We can think about Earth's energy in terms of a budget. Just like a family budget or a government budget, in any particular time period, a certain amount comes in and a certain amount goes out. A family or a company or a government can borrow money so they can send out more than they take in. Earth is different. Earth has a balanced energy budget.

The amount of energy that flows out as heat from Earth's surface and atmosphere to outer space is exactly equal to the amount of energy that reaches the surface and atmosphere. As we have seen, solar radiation accounts for the vast majority of that energy. A much smaller, but very important, amount comes from the interior. Of course, at any given moment more energy could be flowing in than is flowing out. But over the course of a year or longer, this balances and the amount of energy flowing out equals the amount of energy flowing in.

Earth's "matter budget" would look very different. Essentially nothing comes in and nothing goes out. The same stuff keeps getting used over and over. In comparing matter and energy, we say that Earth is a closed system for matter and an open system for energy.

The greenhouse effect adds an important feature to Earth's energy budget. Certain gases in the atmosphere (most importantly, water vapor and carbon dioxide) slow down the rate at which heat escapes from the Earth system. In effect, these gases make the heat stay longer within the Earth system.

People often mistakenly think that the greenhouse effect is a bad thing, that it is something that humans cause. Earth's greenhouse effect has helped make temperatures on the planet comfortable for life for billions of years. It started long before anything resembling humans appeared on the scene.

However, you can always have too much of a good thing. As we learned with the carbon cycle, we are currently causing the amount of carbon dioxide in the atmosphere to increase. By adding greenhouse gases to the atmosphere, we are changing Earth's energy budget. We are making the heat energy stay in the Earth system even longer than it would have. This is the issue of global climate change that we will investigate in Chapter 5.

The main reason we care about Earth's energy budget and the planet's climate is that we and many other creatures live here. The next chapter explores the system of life on planet Earth, including us.

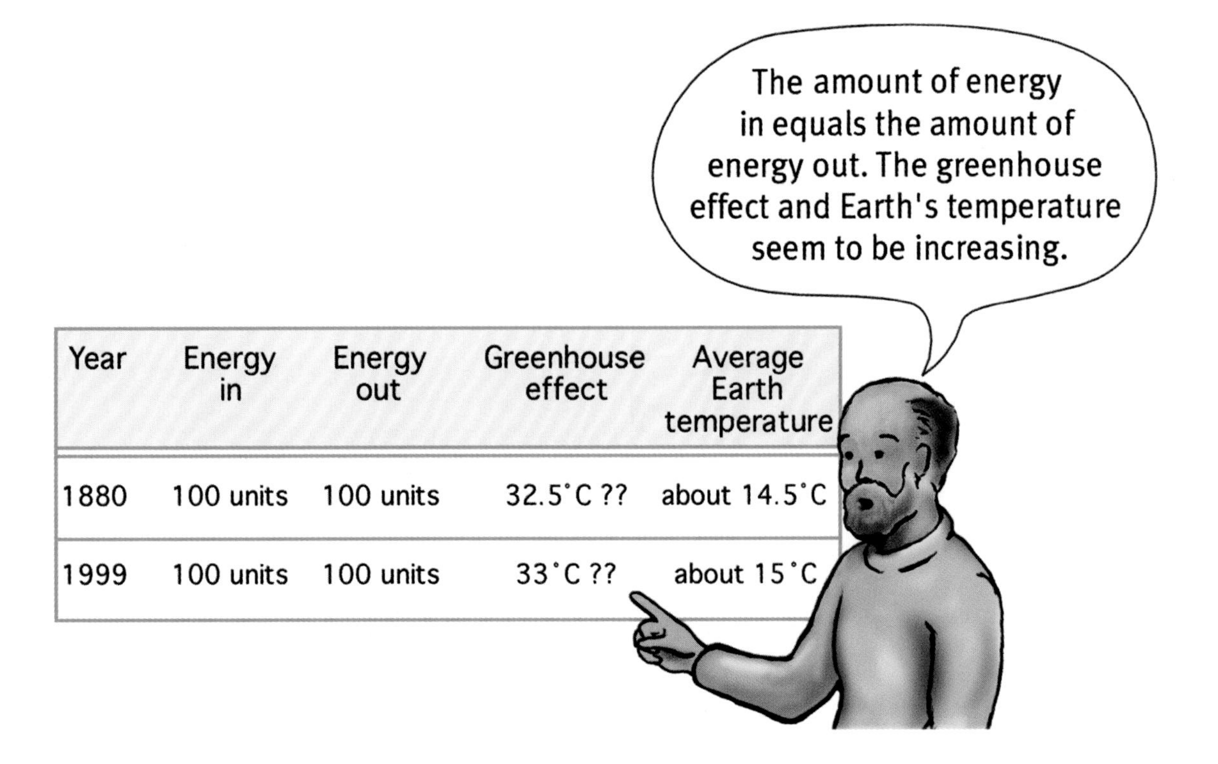

Year	Energy in	Energy out	Greenhouse effect	Average Earth temperature
1880	100 units	100 units	32.5˚C ??	about 14.5˚C
1999	100 units	100 units	33˚C ??	about 15˚C

Explore Chapter 3 on the web...

ENERGY

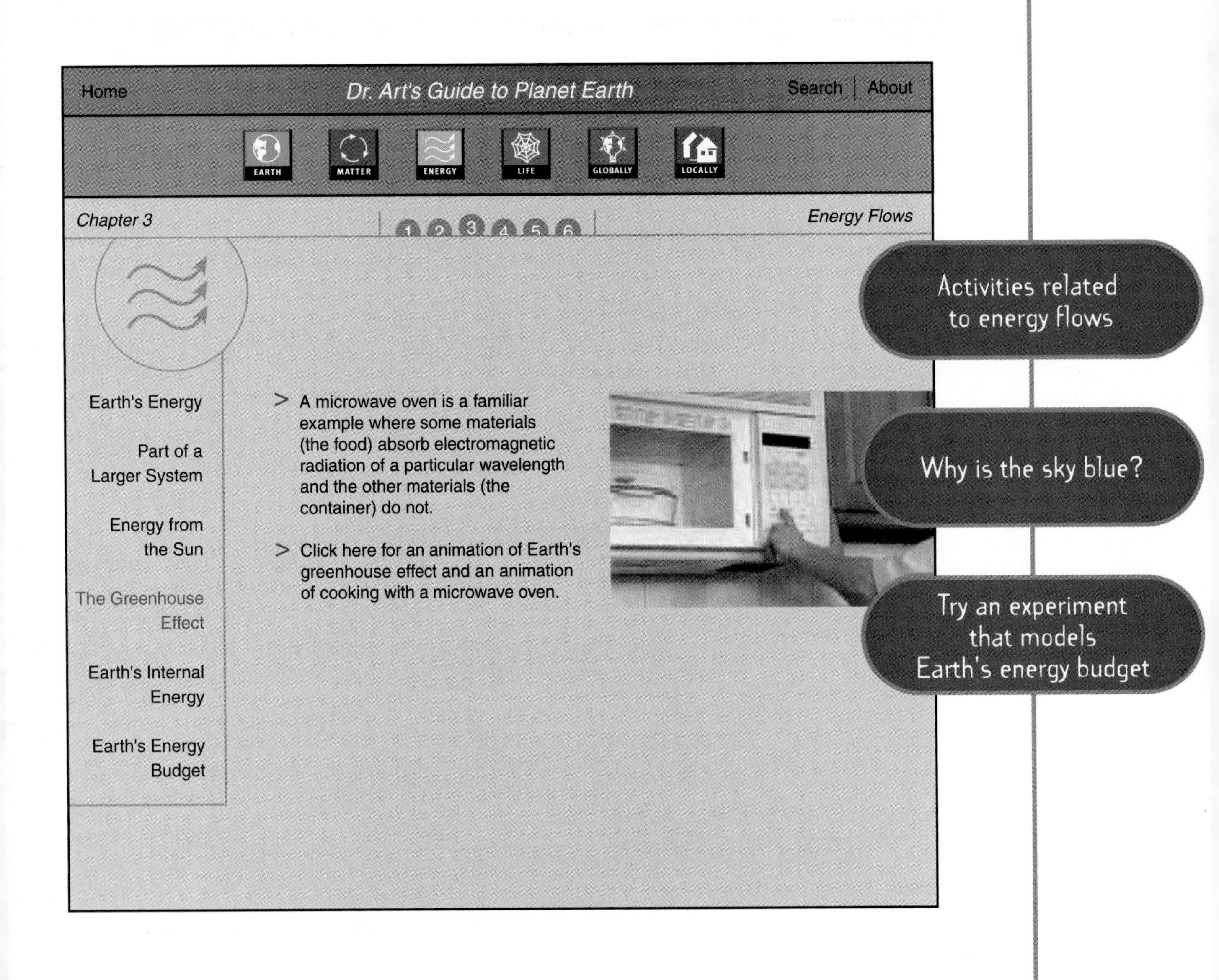

Activities related to energy flows

Why is the sky blue?

Try an experiment that models Earth's energy budget

www.planetguide.net

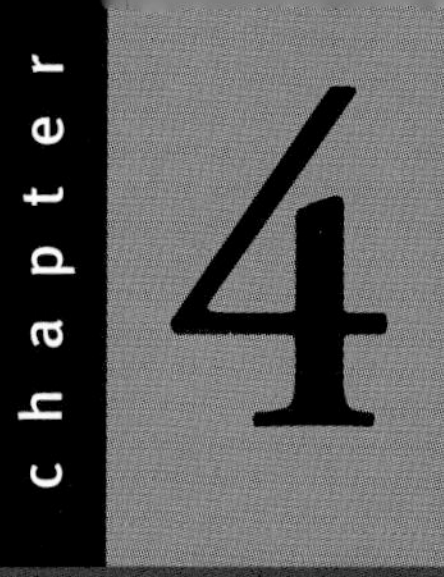

life webs

A Living Planet

Watching Earth Breathe

Who is in the Web?

Ecosystems and Feedback Loops

Shredding the Web

a living planet

Back in the 1960s, NASA hired British scientist James Lovelock to design space flight instruments to test whether Mars has life. NASA chose Lovelock because he had already invented sensitive instruments that could detect tiny amounts of chemicals in our atmosphere. Lovelock thought about the problem and told NASA that he already knew the answer.

DID YOU KNOW?

Earth's atmosphere stands out like a green thumb.

Basically, he compared what we already knew about the atmospheres of different planets such as Earth, Mars and Venus. In the other planets except Earth, the atmosphere is exactly what you would predict from the lifeless laws of chemistry and physics. Earth's atmosphere stands out like a green thumb.

Our atmosphere has way too much oxygen. That oxygen should combine with iron and other chemicals in the Earth's surface and disappear from the atmosphere. In addition, our atmosphere has traces of methane, another chemical that readily reacts with oxygen (methane combines with oxygen to form carbon dioxide and water). The only way our atmosphere could have so much oxygen and also have significant amounts of methane is if something beyond lifeless chemistry and physics were making both oxygen and methane.That something is the network of life on this planet.

Looking at the substances in Mars' atmosphere, Lovelock saw no evidence for life. Organisms living on the surface of a planet will use that planet's atmosphere as a source of the chemicals that it needs and as a place to release chemicals that it produces. The boring chemistry of Mars' atmosphere told Lovelock that Mars is essentially lifeless today.

As far as we know, Earth is unique in the solar system in being a planet of life. Living creatures have dwelled here for almost four billion years. Life has integrated itself so deeply within the operating system of our planet that Earth without life simply would not be Earth.

When we investigated Earth's matter, we asked the first systems question: "What are the parts of the system?" In looking at Earth's energy, we focused on the third systems question: "How is the system itself part of larger systems?" In exploring the system of life on Earth, we are going to focus on the second system question: "How does the system function as a whole?" What is this web of life and how does it work?

LIFE

watching earth breathe

Earth's atmosphere did not always have this unusual chemistry. For its first three billion years, it scarcely had any oxygen. Where did the oxygen come from?

During the first three billion years, bacteria were the only organisms that inhabited the planet. Fairly early in their history, they had already invented a way to capture energy from sunlight and package that energy in chemical form as sugars. As far as living creatures are concerned, this is probably the single most important chemical reaction. We call it photosynthesis, meaning "putting together with light."

In photosynthesis, carbon dioxide combines with water to form sugar and oxygen. The oxygen comes from splitting water (H_2O) and is a by-product of the reaction. Organisms that perform photosynthesis, ranging from single celled bacteria to giant redwood trees, capture the energy from sunlight, package it in chemical form as sugars and release oxygen into the atmosphere. They take carbon dioxide from the atmosphere and pump oxygen into it.

So, doesn't that mean all the carbon dioxide should disappear from the atmosphere and the oxygen should keep increasing? No way. Animals and decomposers help get the carbon from the plant material back into the atmosphere. To get energy, organisms (plants, bacteria, animals and fungi) internally burn sugars back into carbon dioxide gas. This reaction, the opposite of photosynthesis, is called respiration. Organisms release the stored chemical energy by combining sugars with oxygen to form carbon dioxide and water.

Although we did not use the words, we have already encountered photosynthesis and respiration. One part of the carbon cycle illustration from Chapter 2, reproduced here, shows two arrows connecting Land Biomass with the Atmosphere. The arrow pointing downwards in the drawing on page 62 represents photosynthesis – plants and trees taking more than 100 billion tons of carbon out of the atmosphere each year and converting it into sugars. The solid arrow pointing upwards represents respiration – these plants, decomposers and animals internally burning that sugar carbon and converting it back into carbon dioxide.

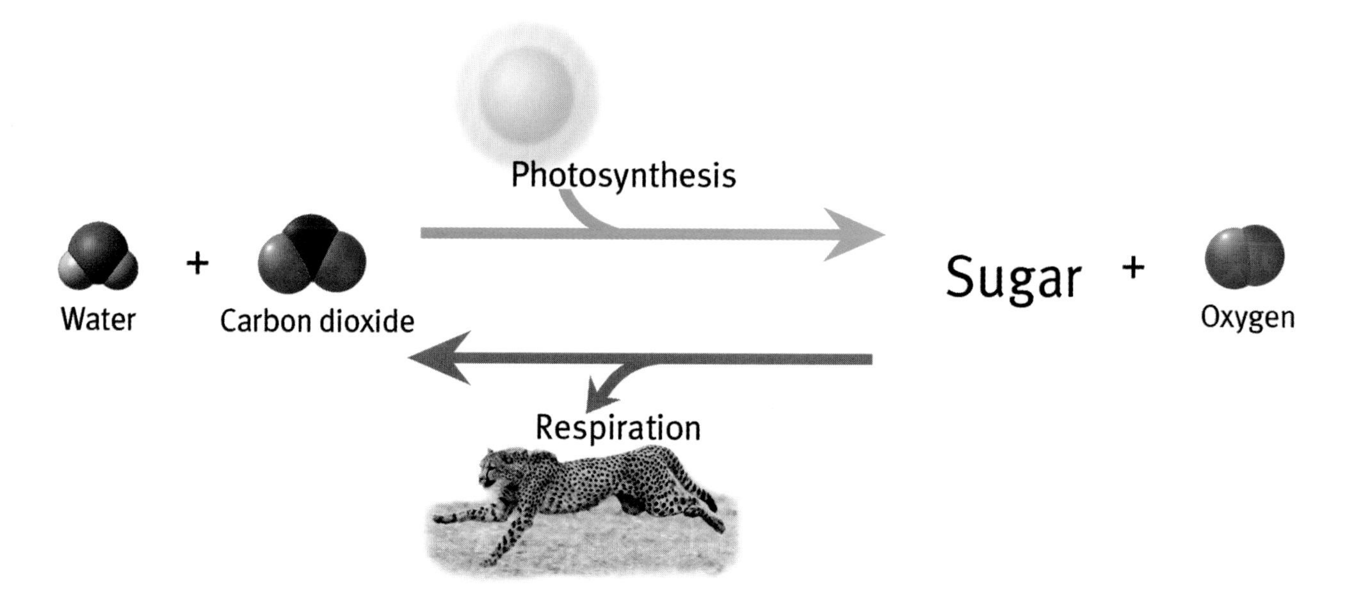

Of course you remember the other parts of the carbon cycle illustration, especially how we are adding carbon to the atmosphere by burning fossil fuels and forests. Back in the 1950s, scientists and government officials realized that we needed to accurately measure the amount of CO_2 in the atmosphere to discover if it was changing. The most famous measuring station, established on the highest mountain of Hawaii's Big Island, has recorded data since 1958.

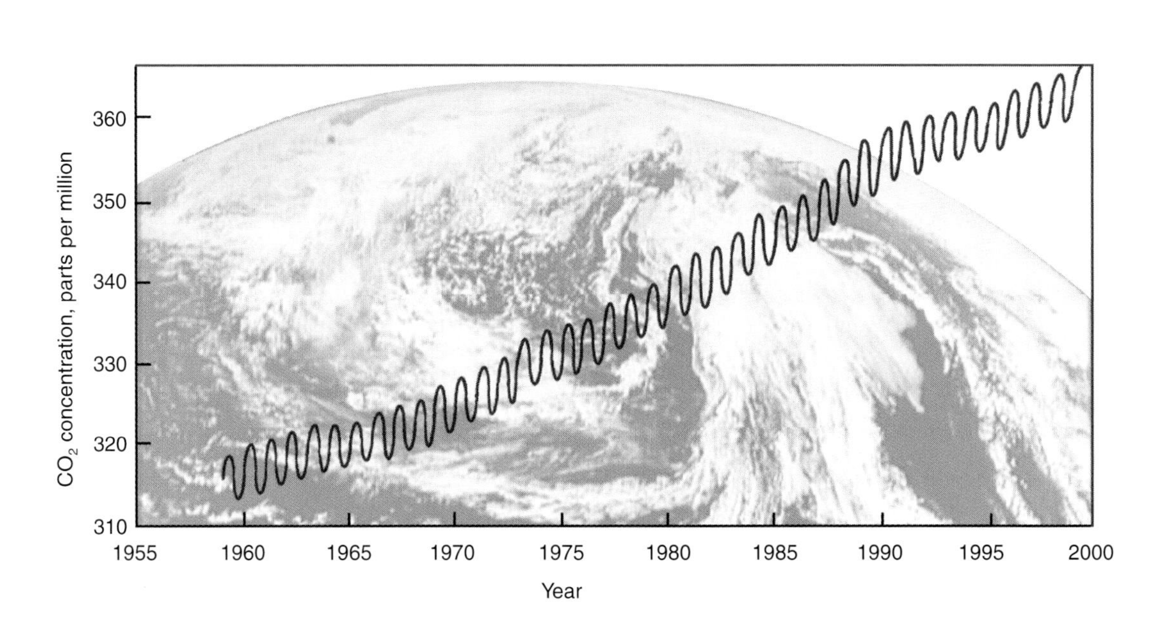

First, notice that the amount of CO_2 in the atmosphere has increased from 316 ppm (parts per million) in 1959 to 364 ppm in 1997. Even way back in the 1950s, we had already been destroying forests and burning significant amounts of coal and oil for more than a hundred years. To have a better idea of human impacts, we would like to know the CO_2 level before the industrial revolution.

Scientists can actually measure CO_2 levels in Earth's atmosphere hundreds and thousands of years ago. No, they don't physically travel back in time. They analyze air bubbles trapped in ice below Earth's surface. The deeper below the surface, the further back in time they can measure. Using this technique, we have data showing that the atmospheric concentration of CO_2 was about 280 ppm in the year 1750 and had stayed fairly constant for the previous 10,000 years. The 1997 concentration of 365 ppm provides strong evidence that human activities have already caused more than a 25% increase in atmospheric CO_2.

What causes the squiggles, the repeating up and down lines, in the graph? Scientists chose the mountain in Hawaii in the middle of the Pacific Ocean so the measurements would be isolated from any major local pollution. They were puzzled by the squiggles, tested various hypotheses and finally concluded that these lines provide a picture of Earth "breathing."

The Hawaii station measures air from the Northern Hemisphere. In the summer of each year, plants strongly increase their rate of photosynthesis absorbing CO_2 from the atmosphere and converting it into sugars. As a result, CO_2 levels go down. When winter comes, photosynthesis decreases while respiration continues releasing CO_2. As a result, CO_2 levels increase during the winter above the summer level. Each year, the CO_2 level starts going down in the spring, reaches a low in late summer or early fall, rises with the beginning of cold weather and reaches a high point before the following spring/summer begins again. The graph shows us Earth's web of life breathing in CO_2 and exhaling it over the course of a year.

Each of us is part of that web of life, converting plant sugars and starch back into atmospheric CO_2. On a sunny day, I like to walk in a park, meadow or forest and become aware that I am breathing with the plant life around me. I inhale the oxygen that plants are releasing at that very moment through photosynthesis. I exhale the carbon dioxide that they inhale at that very moment through their leaves and use in photosynthesis.

who is in the web?

Earth has four kinds of stuff: solid, liquid, gas and living. In comparison with Earth's living matter, there is four thousand times more gas, one million times more liquid and four billion times more solid Earth material. Yet, despite its small amount, life plays very important roles on our planet.

BIG IDEA

We know the number of atoms in the universe better than we know the number of species on our planet.

With respect to mass, almost all of Earth's living stuff is in the form of plant matter. All animal life adds up to only 1% of Earth's biomass. Trees and decaying plant matter account for almost all the mass of Earth's living stuff.

The closer you get to the equator, the more trees there are. In addition to having a warmer climate, there is a lot more land near the equator than near the poles. As a result, tropical forests account for about 40% of Earth's total biomass. This is one reason why many people are concerned about the high rate at which tropical forests are being destroyed. If we burned all of Earth's trees, that would double the amount of carbon in the atmosphere.

Another important way to understand Earth's life is in term of the kinds of organisms that exist rather than their mass. The word "biodiversity" refers to the number and kinds of different organisms. We know very little about Earth's biodiversity. We have a more accurate scientific estimate of the number of atoms in the universe than we do of the number of different species on our planet.

Scientists have currently named and catalogued about 1,500,000 different species of organisms. The estimates of the total number range from 5 million to 30 million or even more. Of the 1.5 million that have been described, all that we know about the vast majority of these is what they look like and where a few specimens were obtained.

Where is this biodiversity located? Here, too, the tropical forests play a prominent role. Biologist E. O. Wilson once found as much diversity of ants on one tree in Peru as exists in all the British Isles. A naturalist in 1875 described 700 species of butterflies within an hour walk of an Amazon river town while all of Europe has only 321 different butterfly species. An area in Indonesia totaling approximately 25 acres contained as many different tree species as are native to all of North America. This wealth of biodiversity may disappear in smoke before we even know what we have lost forever.

ecosystems and feedback loops

Having discussed how much living matter there is and how many different kinds there may be, we are still missing a very important part of understanding life on Earth. How is it organized?

BIG IDEA

All the different ecosystems have a similar pattern of organization.

The millions of species occupy specific places. The scientific term ecosystem refers to the organisms that live in a particular place, their relationships with each other and their interactions with the physical and chemical parts of their environment. You may have experienced different ecosystems such as a lake, meadow, creek, forest, coastal tide-pool, coral reef or desert.

Ecosystems, like other systems, can be described or investigated at many different levels. There are ecosystems within ecosystems within ecosystems. A meadow ecosystem includes plants, insects, gophers, snakes, deer, fungi and bacteria. The forest in which the meadow is located is another ecosystem. An even larger ecosystem would be a mountain containing the forest, meadow and perhaps a lake. Earth's web of life is the sum total of all Earth's ecosystems.

All the different ecosystems have a similar pattern of organization. They all require a source of energy and a group of organisms that can capture that energy and store it in chemical form. For the vast majority of ecosystems, the sun provides the energy. Plant life, ranging from microscopic algae to towering redwood trees, capture the energy in sunlight and store it as chemical energy in sugar molecules.

The organisms in an ecosystem that capture the energy are called producers (labeled P in the illustration below). All the other organisms in the ecosystem depend either directly or indirectly on the producers for their food.

Animals are consumers, either eating plants (H, herbivores) or other animals (C, carnivores). Another group of consuming organisms breaks down dead plants and animals (D, decomposers).

In any ecosystem, the producers, consumers and decomposers establish a network of feeding relationships called a food web. The same material keeps getting used over and over again as one organism eats another and they all decompose. Recycling is the ecosystem way of life.

Analyzing the energy flow through the ecosystem provides another perspective on its organization. The organisms that have the highest total energy flowing through them are the producers. All the biological energy that flows through the ecosystem's organisms must first be captured by these producers. In the course of living and reproducing, some of that energy escapes as heat to the atmosphere. The herbivores (cows, sheep, squirrels, etc.) that eat the producers spend a lot of energy in maintaining their body temperature, mating, eating and protecting themselves. This energy eventually escapes to the atmosphere as heat. Therefore there is less biological energy available to support the carnivores (snakes, owls, mountain lions, people) that eat the herbivores.

One result is that a typical terrestrial ecosystem will have five to ten times as much biomass in plant life than it will in herbivores. It will also support five to ten times as much biomass in herbivores as it will in carnivores. This feature is often portrayed as a pyramid showing the producers as the broad ecosystem base with a narrower middle representing the herbivores and a very narrow top representing the carnivores.

BIG IDEA

Recycling is the ecosystem way of life.

Think about the forest/meadow ecosystem shown on the previous pages. Imagine that a new disease kills all the mice. How might that affect the rabbit population?

The rabbits might increase since there may be more plant food for them to eat. On the other hand, the owls and foxes might eat more rabbits to replace the missing mice. This could cause a decrease in the rabbit population.

This type of question often occurs when we study a system. What happens when one of the parts changes? How do the parts connect with and influence each other?

In general, the parts of a system connect with and influence each other in two different ways. We call these balancing feedback loops and reinforcing feedback loops.

A **balancing feedback loop** – surprise! – tends to keep things in balance. Predators and prey exist in a balancing feedback loop. If a mouse population increases, the hawks tend to increase since they have more mice to eat. The increase in hawks then reduces the mouse population, serving to balance the initial increase in mice.

Balancing feedback loops are very common. A thermostat is an example of a balancing feedback loop. A room gets too cold, triggering the thermostat to turn on the heater. When the room reaches the set temperature, the thermostat turns the heater off. The room temperature stays in balance, moving just a few degrees above and below the set temperature.

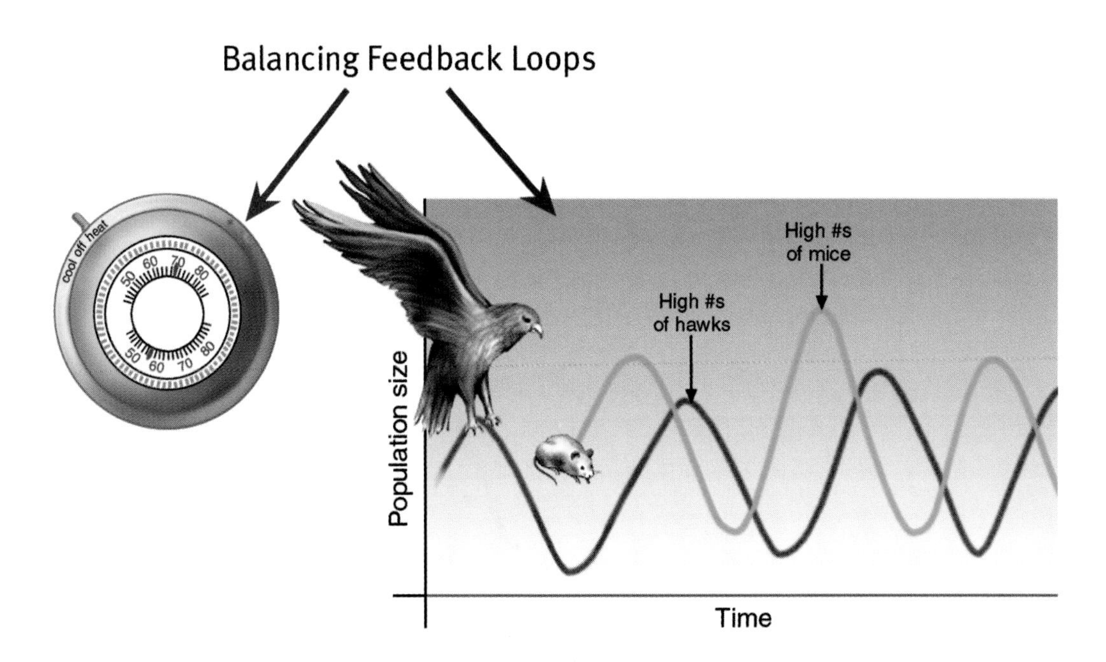

With a **reinforcing feedback loop**, a change in one direction causes more change in the same direction. The high pitch squeal of a microphone is one example of reinforcing feedback*. As another example, think about ten rabbits being brought to a new continent where they have abundant food and no natural predators. Each rabbit on average causes ten new babies so the population quickly becomes 110. Each of these results in ten new rabbits, a population of 110 + 1,100 = 1,210. This reinforcing feedback loop (more rabbits causes more babies causes more rabbits causes more babies) quickly results in a population explosion with millions of rabbits reproducing like rabbits all over Australia.

Complicated systems such as ecosystems or the planet's web of life have many parts that all directly or indirectly connect with each other. A change in one part will cause changes in other parts. Some of the changes get balanced while others are reinforced. All these influences feed into each other causing the whole system to change, often in unexpected ways. You have probably experienced situations where simple actions cause unexpected results.

* The microphone picks up some electronic noise, feeds it into the amplifier which makes the sound louder and then broadcasts it via the speakers. The microphone picks up the louder noise, feeds it back into the amplifier and this loop repeats making the sound louder and louder.

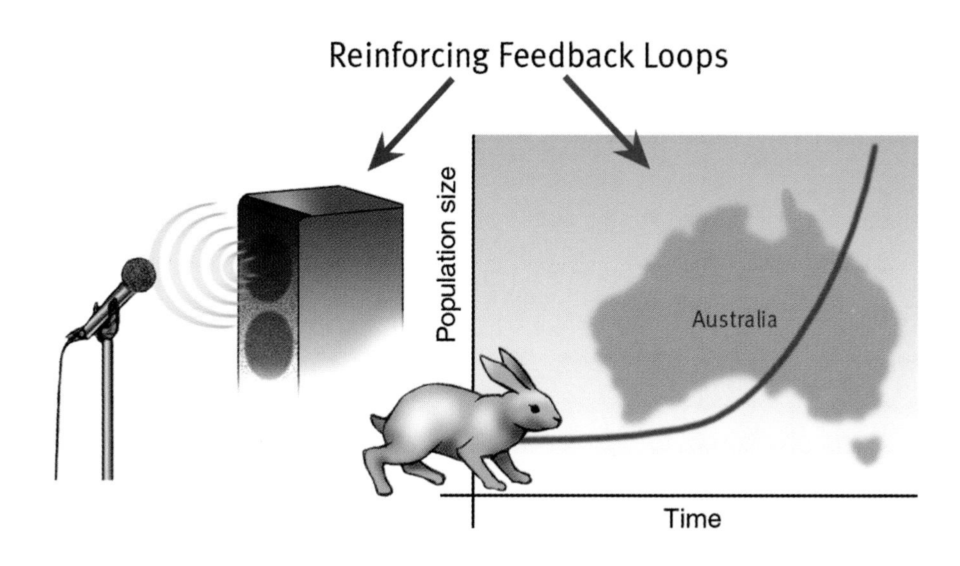

"Parachuting Cats Into Borneo" is a famous example. The World Health Organization (WHO) sprayed the insecticide DDT in Borneo in the 1950s in order to fight malaria, a disease spread by mosquitoes. The people lived in homes with thatched roofs. Suddenly, their roofs collapsed.

In addition to killing mosquitoes, the DDT had killed parasitic wasps that preyed on caterpillars that ate the roof materials. Without the wasps, the caterpillars multiplied out of control and destroyed the roofs. The local gecko lizards also died from eating DDT-poisoned insects. The dying geckos were caught and eaten by house cats that then died from the DDT. The death of the cats caused an increase in rats which threatened to cause an outbreak of bubonic plague (a terrible disease which is spread by rats). WHO then parachuted cats into Borneo to try to control the rat population. They clearly did not have this in mind when they began spraying DDT.

We keep learning this lesson about the web of life. All the parts are connected via feedback loops. When we change the web of life, it is hard to predict the consequences.

LIFE

shredding the web

From our earliest beginnings, humans have affected the web of life. Since everything is connected, we could say the same thing about any organism. The difference is that we now have a large human population and powerful technologies that have far-reaching effects. Scientists estimate that we currently take about one-third of the photosynthesis energy captured and stored by Earth's land plants.

BIG IDEA

We are harming the planetary web of life in at least six different ways.

We are harming ecosystems locally and globally in at least six different ways (see next page). As a result of all these actions, we are beginning to shred the existing web of life. Should we care? Can we do anything about it? In the next chapter, we will analyze three global environmental issues, beginning with our effects on Earth's biodiversity. The three Earth Systems principles (Matter Cycles, Energy Flows, Life Webs) will help us understand these global environmental issues as well as the local environmental issues that we will explore in the final chapter.

HABITAT FRAGMENTATION
Isolating patches of natural habitat

HABITAT DESTRUCTION
Physically destroying natural habitat

POLLUTION
Adding chemicals to natural habitat

EXCESSIVE HARVESTING
Logging, fishing and hunting at faster rates than nature can replace

EXOTIC SPECIES
Introducing plants and animals into new ecosystems where they grow out of control

CLIMATE CHANGE
Increasing the amount of greenhouse gases in the atmosphere resulting in changes to Earth's climate

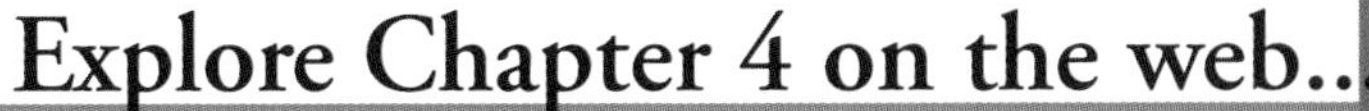

Explore Chapter 4 on the web...

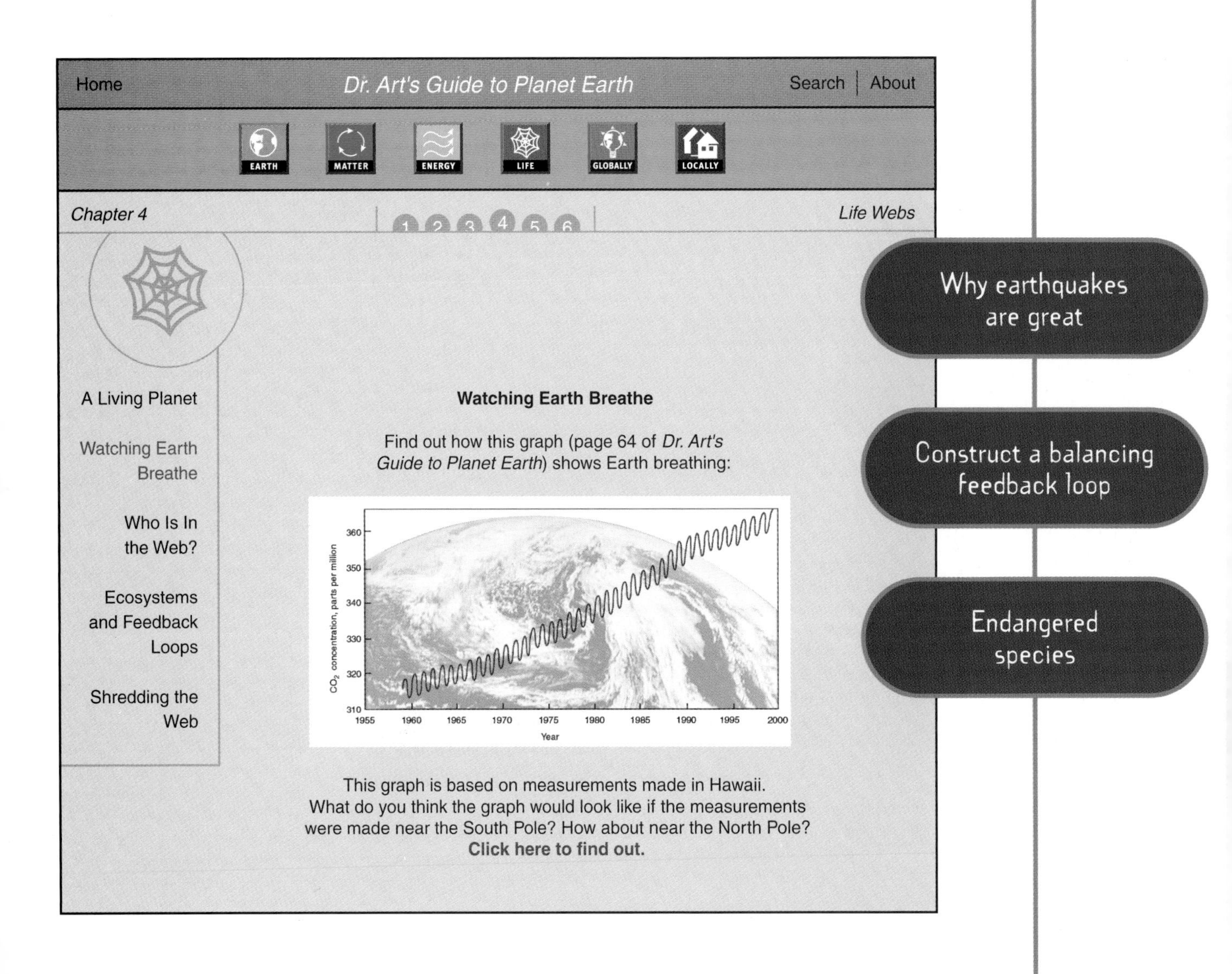

www.planetguide.net

chapter 5

think globally

Save the Planet?

Extinction

The Ozone Layer

Climate Change

Ice Age or Hot House?

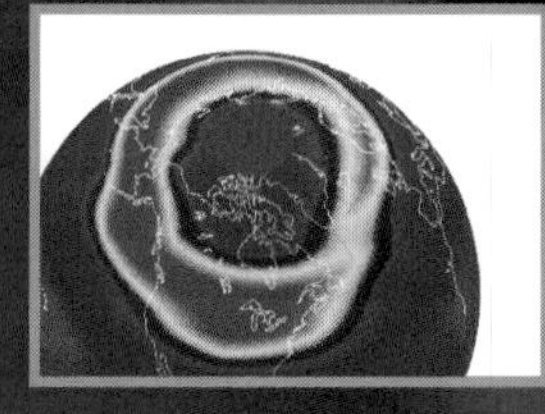

save the planet?

You probably have seen the phrase, "Save the Planet." My advice? Don't worry about saving planet Earth.

Our planet has lasted more than four billion years and survived far greater calamities than anything we can do. We cannot destroy planet Earth. Fortunately, we cannot even destroy life on our planet.

About 65 million years ago, a large asteroid or comet probably crashed into Earth. The evidence indicates that the force of the impact was equal to exploding 7,000 times the amount of all the world's nuclear weapons. Even that extreme catastrophe did not destroy all of Earth's life. It probably caused the extinction of the dinosaurs and 75% of all species living at that time.

Does that mean we don't have to worry how our actions can affect the environment? I don't think so. Even though we cannot destroy life on Earth, we can cause changes that would be very harmful to many of Earth's current inhabitants, including ourselves.

I would bet that on any day you could find something in the newspaper, TV or radio that talks about one or more environmental issues. In general, there are two different types of environmental issues -- local and global. The local issues concern the area close to where we live and the things in our environment that affect us every day (food, air, water, garbage). In contrast, the global environmental issues can change conditions throughout the planet.

This chapter explores three issues that can change conditions on a planetary scale. These global environmental issues are:

Extinction

high rates of species extinction and damage to ecosystems

Ozone

destruction of the ozone in the upper atmosphere that protects organisms from the sun's ultraviolet (UV) radiation

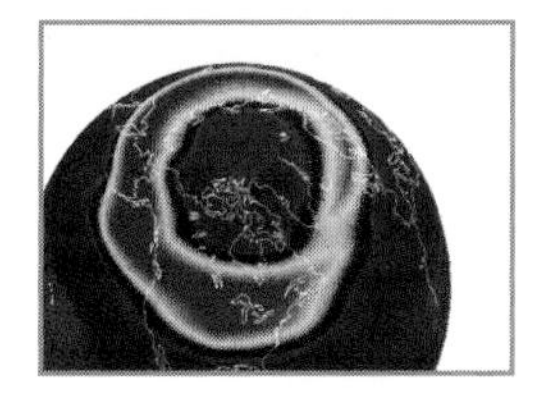

Climate

increase in greenhouse gases in the atmosphere resulting in climate changes throughout the planet

Even without human influences, Earth goes through major changes over time. As we explore these three global environmental issues, we will examine how Earth has changed in the past. How fast did those changes occur? How long does it take for Earth to recover from a major, global catastrophe? How do changes that humans may be causing today compare with previous changes?

extinction

Life on Earth began about 3,800,000,000 (three billion, eight hundred million) years ago. None of us were there to see it. In fact, for most of life's history, the only living things were microbes, single-celled creatures too small for us to see without a magnifying device.

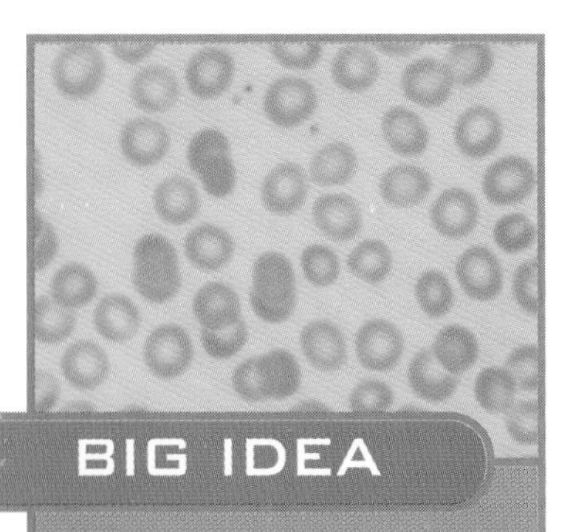

BIG IDEA

For most of Earth's history, the only living things were microbes.

These tiny organisms invented photosynthesis (capturing the sun's energy in chemical form), motility (moving in a direction), behavior (moving towards things they like and away from things they do not like), and sex (sharing genetic information). Through photosynthesis, they put oxygen into our atmosphere. They adapted to many different environments and changed our world.

The first multi-cellular animals evolved about 550 million years ago. These ocean-dwelling organisms did not have backbones but they did have hard shells, some of which have been preserved as fossils that show us what these ancient organisms looked like.

Scientists use the word biodiversity to describe the number of different kinds of organisms, or in other words, the variety of life on Earth. The illustration on the next page shows how Earth's biodiversity has changed over the last six hundred million years. As we would expect, the graph increases from the left to the right, indicating that more different kinds of organisms exist today than when the first multi-cellular animals appeared.

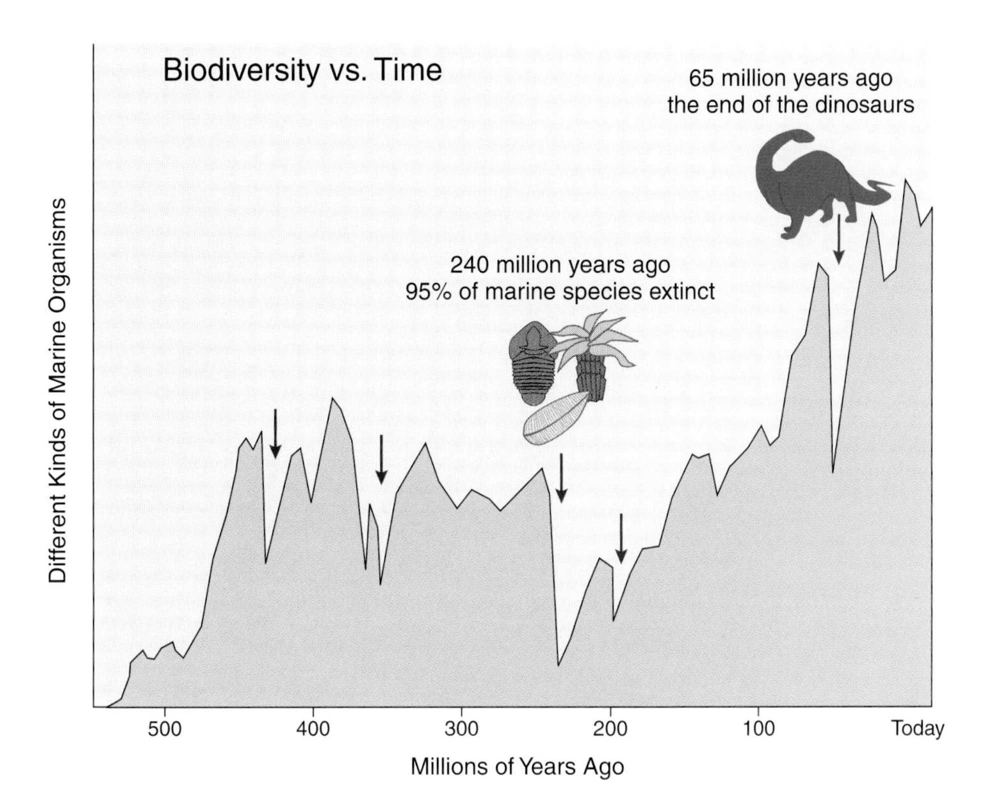

Note that biodiversity has suffered some major setbacks in its long history. During the past 500 million years, there have been five major disruptions that we call mass extinctions. The most extreme occurred about 240 million years ago when about 95% of the then-living marine species disappeared. The most famous mass extinction occurred about 65 million years ago, marked the end of the dinosaurs, and was probably caused by the meteorite collision described earlier.

After each mass extinction, Earth's biodiversity eventually recovered. However, the process takes time, lasting millions of years. Further, it is not the old organisms somehow coming back. Recovery involves new species evolving and replacing the old ones. Extinction is forever.

BIG IDEA

We may be in the middle of a mass extinction.

Extinction occurs even in "normal" times. Biologists estimate that this background level of extinction averages about 10 to 25 species per year. What about today? Many ecologists are concerned that we are already in the midst of mass extinction, this time caused, not by a meteorite from outer space, but by Earth-born creatures that walk on two legs.

Of course you remember "Shredding the Web," the last part of Chapter 4 (p. 74-75). Humans engage in six different types of activities that harm the existing web of life. These include destroying habitat, polluting, and harvesting (hunting, fishing, logging) at faster rates than nature can replace. In many ecosystems, we do all these things at the same time. When people move into or economically develop a new area, they build roads that fragment the habitat, cut down the forests, spill chemicals on the ground and in the rivers, bring in domestic animals, and kill the local animals.

Now we are threatening to change the climate as well. A plant or animal species that has already decreased in numbers due to loss of habitat and exposure to pollutants may not be able to survive a change in climate. If the new climate makes its current habitat unsuitable, it may not simply pack up and move to a new area whose new climate could be satisfactory. First, highways, suburbs and cities may block the way. Second, species depend on each other. The climate in an area may be perfect, but an organism will not be able to live there if that area does not have the plants and animals that it needs for food and shelter.

What is happening to the web of life today? Many respected biologists believe that we are already beginning to experience a mass extinction that is as severe as the mass extinctions that occurred in the past. The normal background extinction rate is about 10 to 25 species per year; the current rate is probably at least several thousands of species per year and may be ten times that high.

How can we continue our normal lives without even being aware of a mass extinction? Well, most of us live in or near cities, far from the areas where most of Earth's biodiversity exists. We live far from the areas that are currently experiencing major habitat destruction, the main cause of extinction today. Tropical rain forests that are home to about half of Earth's biodiversity are being destroyed every day.

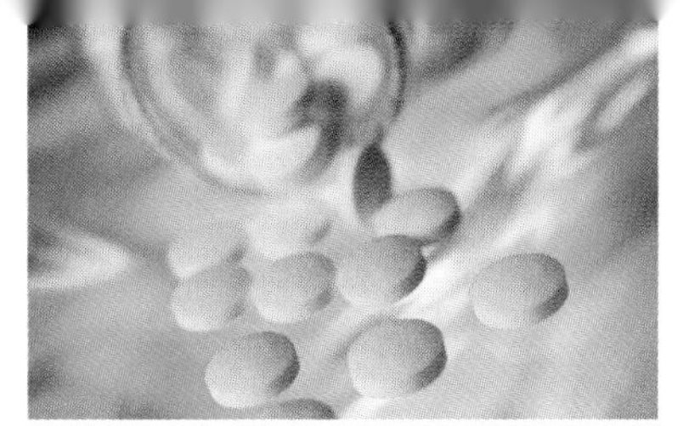

Should we care about all these species disappearing? Most of them are insects and even smaller organisms that none of us would ever see.

Many people want to prevent extinctions because they believe that it is morally wrong to destroy ecosystems and cause other organisms to disappear forever. Many people also believe that the natural world should be protected simply because it is beautiful. Both of these arguments say that we should protect ecosystems even if they do not have any practical, economic importance.

Another kind of argument states that Earth's biodiversity has tremendous economic and practical value, and that we are already destroying irreplaceable wealth. About one quarter of the medical drugs produced in the United States contain ingredients that were first discovered in plants. Aspirin, the most commonly used medical drug, is one example. The rosy periwinkle, a plant which grows only in Madagascar, gave us a drug that cures almost all cases of childhood leukemia, a disease that previously killed almost all its victims.

Plants have developed an incredible variety of chemicals over millions of years. When a new disease or insect attacks a plant crop, scientists search the natural world for varieties that are resistant to that disease or insect. They can then protect important crops such as rice and wheat by breeding in the resistance from the wild varieties. When a plant species become extinct, we may have lost forever a cure for AIDS, cancer, or diseases that attack our food crops.

The natural world also provides free services that we tend to take for granted, including clean air, clean water and food. Organisms play important roles in Earth's cycles of matter such as the carbon, nitrogen and sulfur cycles.

How many species can disappear before today's web of live unravels? The answer is – we do not know. We don't know the details of how most ecosystems work. We don't know how ecosystems interact with each other. We don't know how different ecosystems or combinations of ecosystems support the larger global system. We don't know how many species there are today, how many are going extinct right now, and what will happen if we continue our present activities. We don't know.

BIG IDEA

We Don't Know:
how many species there are, how many are going extinct, what will happen to the web of life.

There is one thing that we probably do know. Humans like to protect cute, exciting, powerful and/or cuddly creatures. We want to save the whales, cheetahs, and pandas. We also like to protect ourselves. However, we and other carnivores sit at the top of ecosystem pyramids. That makes us more vulnerable to ecosystem changes.

The producers, who capture the sun's energy, and the decomposers, who help recycle matter, play crucial roles in ecosystems. These very important parts of Earth's biodiversity are the plants (including plankton, microscopic organisms that support ocean ecosystems), and the ugly, the invisible, and the smelly (including fungi, bacteria and insects). These are creatures that we usually do not see on TV, refrigerator magnets, the zoo, or newspaper articles.

Many scientists and government officials now try to protect ecosystems rather than focusing on an individual species. When a species is endangered, we can take that as a warning sign that we need to protect the ecosystems to which it belongs. That way, we can protect the producers, the smelly, the invisible, and the ugly, and maybe ourselves in the long run as well.

the ozone layer

The second global environmental issue involves the thin but vital layer of ozone in the upper atmosphere. This ozone protects Earth's organisms from the sun's ultraviolet (UV) radiation. Chemicals that humans have produced are destroying this ozone and causing an increase in the amount of UV radiation that reaches the Earth's surface.

Ozone is a form of oxygen. While the familiar form of oxygen contains two oxygen atoms bonded together (O_2), ozone has three oxygen atoms connected with each other (O_3). This change in chemical structure causes these two different forms of oxygen to have different properties. We breathe the two-atom form in order to live. The three-atom form is actually quite toxic to us.

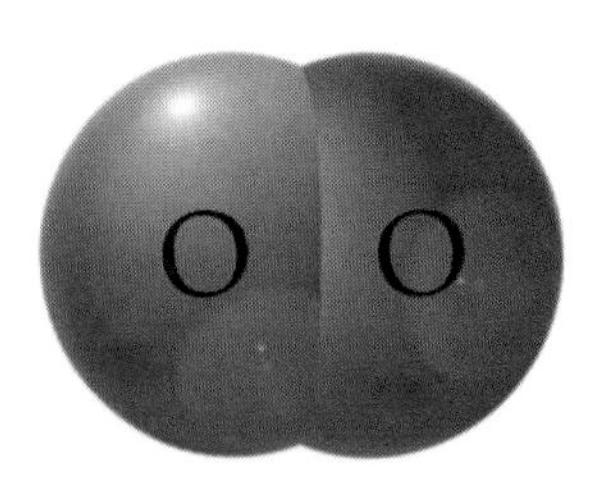

Oxygen (O_2)

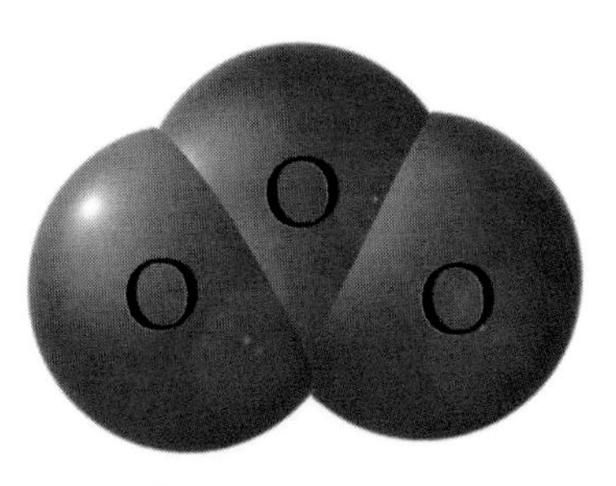

Ozone (O_3)

GOOD OZONE AND BAD OZONE			
TYPE OF OZONE	WHERE IS IT?	HOW IS IT MADE?	WHAT DOES IT DO?
"Good" ozone	Upper Atmosphere	Natural result of oxygen reacting with UV light	Protects life from the sun's UV rays
"Bad" ozone	City smog	Results from pollutants (e.g., car exhaust) reacting with sunlight	Causes health problems, especially with breathing

Fortunately, most of Earth's ozone is in the upper atmosphere, 15 to 50 kilometers (about 9 to 30 miles) above our heads. Up there, it absorbs the sun's UV radiation and protects us. Actually, some ozone occurs in the lower atmosphere that we breathe. This ozone is part of the city smog caused by pollution, and it is a local environmental issue because it harms our lungs. Some people call them good ozone and bad ozone.

We care about the good ozone because it protects life from the sun's UV radiation. Even small increases in UV exposure can cause increases in skin cancer and eye cataracts, and perhaps damage to the immune system. We don't know how much different levels of UV radiation will damage Earth's incredible variety of interlinked organisms and ecosystems. At the least, it would be more stress on the web of life.

If we traveled back in time, we would discover that Earth's atmosphere did not always contain ozone. In fact, up until about 2 billion years ago, Earth's atmosphere had essentially no oxygen. Remember that for the first two billion years of life, the only organisms were microbes living in the ocean who were busy inventing swimming, behavior, sex, and . . . . photosynthesis.

Well, little did they know it, but the oxygen that they kept making through photosynthesis eventually built up to the point where it changed the atmosphere and the history of life on Earth. As soon as there was oxygen in the atmosphere, it reacted with incoming UV rays and formed ozone. This ozone absorbs UV radiation and prevents it from reaching the lower atmosphere and the Earth's surface. The stage was set for life to eventually move out of the protective ocean and onto the land.

Using the fast-forward button on our time machine, we return to the present and discover a newspaper headline saying something about the ozone hole. Why would people make chemicals that destroy the protective ozone in our upper atmosphere?

BIG QUESTION

Why would people make chemicals that damage the Earth's ozone layer?

They were trying to do a good thing. In the early 1900s, chemists were trying to create the ideal chemicals to be used in refrigerators. These substances had to be stable so they did not break down right away. They also needed to be chemically inert, meaning that they would not interact with other substances. That way they would not corrode the refrigerators or cause health problems for humans.

The chemists succeeded, resulting in a huge new industry for refrigerators and air conditioners. Most people in the industrial world now take it for granted that we can have safe, cold food as well as comfortable indoor temperatures during the summer. The chemicals that made it possible are small carbon-based molecules that also contain chlorine and fluorine atoms. They are called chlorofluorocarbons, CFCs for short.

Meet a chloroflurocarbon (CFC)

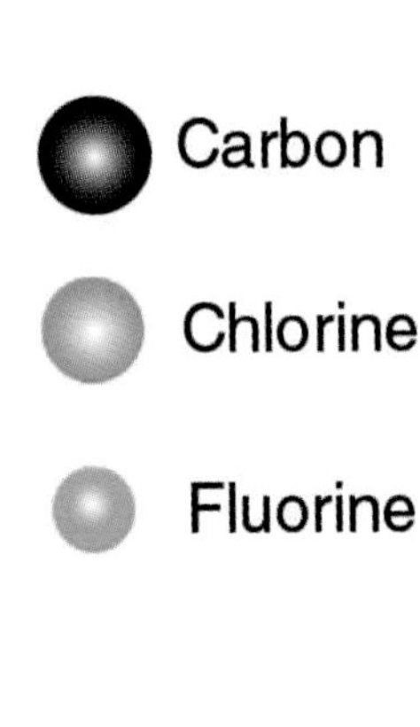

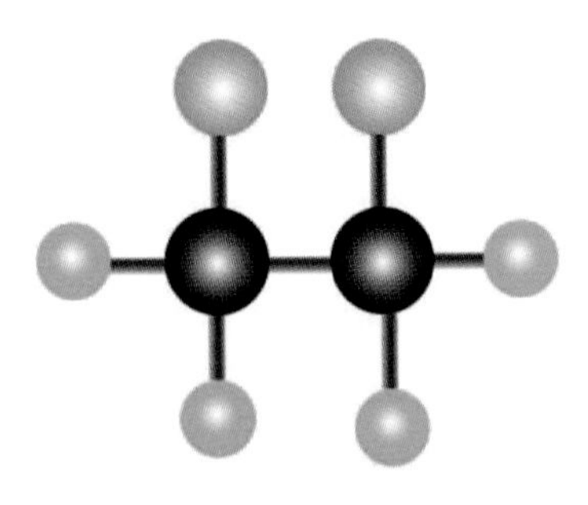

Well, guess what happens to a chemical that is very stable and that does not interact with other substances? Nothing. It keeps accumulating. It does not participate in Earth's cycles of matter. As we find more uses for it and make more of it, the more the chemical eventually finds its way into the atmosphere and just hangs around.

It hangs around outside the cycles of matter . . . until it drifts miles above the Earth and reaches the upper atmosphere, where ozone molecules and UV radiation interact. There it finally meets something that can tear it apart – high energy UV radiation. When a UV ray breaks up a CFC molecule, it releases chlorine. Unfortunately, chlorine atoms destroy ozone. Each of these CFC chlorine atoms that is released in the upper atmosphere can destroy 100,000 ozone molecules.

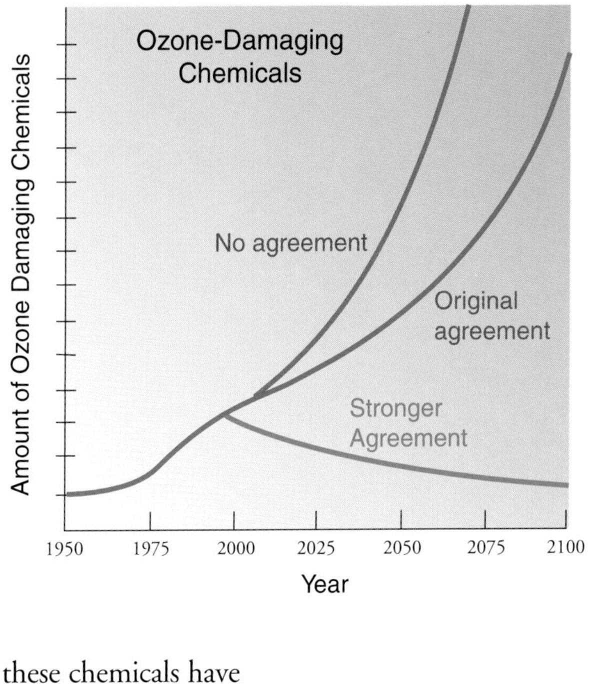

Scientists and environmentalists had been concerned about the ozone layer, but businesses and governments generally opposed making changes until there was more proof. After the discovery of the ozone hole, scientific research provided very strong evidence that CFCs cause the hole. With that information, the world community organized and took action.

Environmental ministers from 24 nations, representing most of the industrialized world, met in Montreal in 1987 and agreed to limit the production of substances that damage the ozone layer. A stronger agreement in 1990 provided bigger and quicker reductions in the use of these chemicals. The graph shows how much these chemicals have already increased in the atmosphere and the predictions of what will happen in the future.

We know that damage to the ozone layer increases the amounts of UV light that reaches the surface. Since the ozone levels naturally change over time due to weather conditions, volcanic activity and other causes, we do not have accurate information about how much extra UV radiation is happening right now because of CFCs and other manufactured chemicals. We do know that the protecting ozone layer has decreased over populated areas, resulting in increased UV radiation at some times of the year. We currently expect the ozone layer to slowly recover and return to its pre-industrial level between the years 2050 and 2100.

BIG IDEA

Unpleasant surprises happen.

This ozone story tells us that very unpleasant surprises can happen if we ignore Earth's cycles of matter. We manufactured large amounts of a new kind of chemical. Since they could not be naturally recycled, they accumulated in the atmosphere. Eventually, they harmed Earth's ozone layer. We are beginning to realize that comparatively small amounts of man-made chemicals can dramatically change important features of the Earth system. Fortunately, we probably have caught this problem before it could become a global disaster.

The Ozone Hole Surprise

Scientists knew that we had to be concerned about the ozone layer in the upper atmosphere. After all, it is pretty thin. If all the ozone up there were concentrated and brought to ground level, it would form a shell around the Earth that is only 0.3 cm wide. In reality, the ozone is spread very thinly in the stratosphere from 15 to 50 kilometers above the ground.

So scientists were measuring ozone levels all around the globe. A British group in the early 1980s measured that the ozone dramatically fell during the Antarctic Spring months (Fall in the Northern Hemisphere). At first they were very reluctant to publish their results. After all, they were using old equipment. The Americans using sophisticated equipment on a NASA satellite were reporting normal ozone levels. Finally, they published their results in a major scientific journal.

Suddenly lots of people were talking about the "ozone hole." Actually, it is not a hole. It is a very large area over the South Pole where the ozone concentrations drop 60% or more during the Spring months. The area affected is as large as the United States plus half of Canada.

How did the NASA scientists miss the ozone hole? Their data showed the same hole. Unfortunately, they had programmed their computer to ignore any data which was too far outside the range that they had expected. One of the important lessons to learn in science (and in life) is to be prepared for surprises, both pleasant and unpleasant.

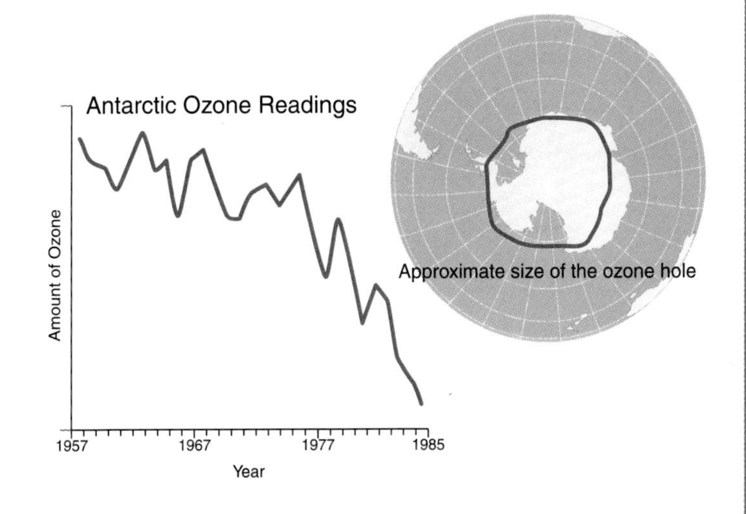

climate change

We have already met the third global environmental issue: climate change. Our three Earth systems principles help us understand this issue. We are disturbing the cycles of matter by putting greenhouse gases into the atmosphere. These gases interfere with the planet's flows of energy. The resulting climate change can damage the web of life.

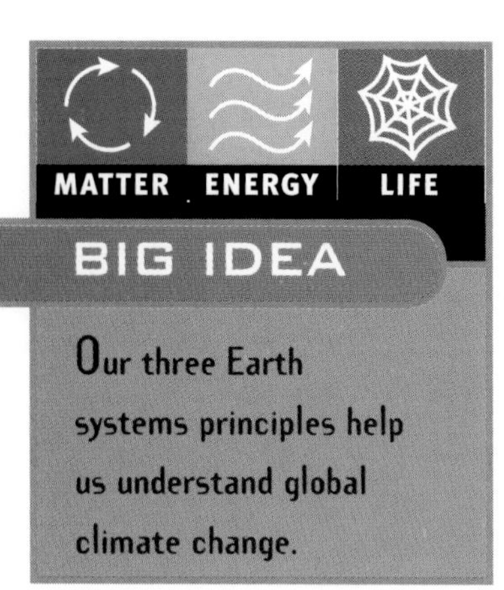

Climate is different than weather. When we talk about weather, we care if it is raining, sunny, hot or cold in some particular place today or next week. With climate, we care about the pattern of weather over a longer period of time and usually over a broader area. Global climate refers to the pattern of temperatures and precipitation for the planet as a whole.

The Major Cold and Warm Periods During Earth's History

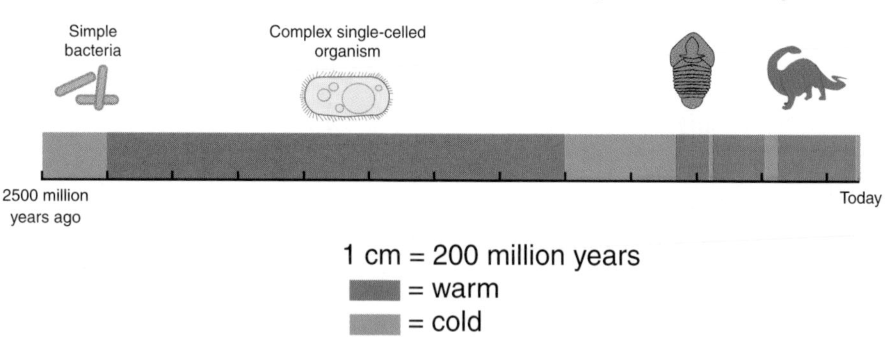

In Earth's fiery beginnings, it was so hot that it completely melted. Once it settled down about 4 billion years ago, Earth has never been so hot that the oceans boiled or so cold that the oceans completely froze. During those 4 billion years, Earth has been mostly warm with occasional cold periods. Over the past 2,500,000,000 years Earth has been warm 75% of the time and cold about 25% of the time.

When it is warm, Earth has little or no permanent ice covering its land. That may be a surprise to most of us who are used to thinking of Earth with permanent ice at both poles. During its cold periods, Earth has lots of ice that covers land throughout the year. Today about 10% of Earth's land surface is covered with ice. 20,000 years ago, ice covered about 30% of Earth's land surface.

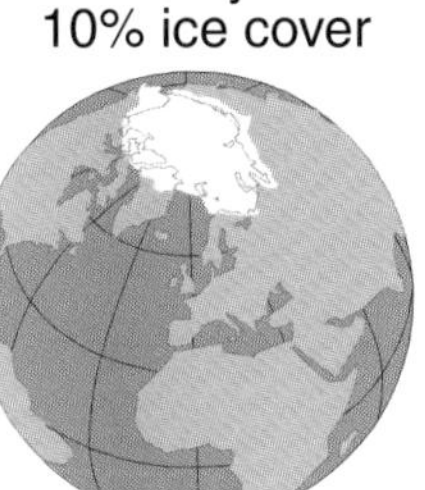

Earth is actually considered to still be in a cold period that has lasted about two and a half million years. Right now, we are in what is called an interglacial, a warm part of that cold period. We came out of a deeper Ice Age only about 10,000 years ago.

What causes these changes in Earth's climate? Three factors determine whether Earth has a cold or warm climate:

- How much solar energy enters the Earth system;
- How solar energy circulates within the Earth system;
- How heat energy leaves the Earth system

Solar Energy Into the Earth System

The sun provides the vast majority of Earth's energy. Earth's orbit around the sun changes in a number of ways. For example, the tilt of Earth's axis changes from 21.5 degrees to 24.5 degrees in a 41,000 year cycle. A different 100,000 year cycle involves a change from a nearly circular orbit to a more elliptical one. During the past several million years, these changes appear to cause a repeating pattern of long cold periods with briefer warm periods (the interglacials) that last ten to twenty thousand years.

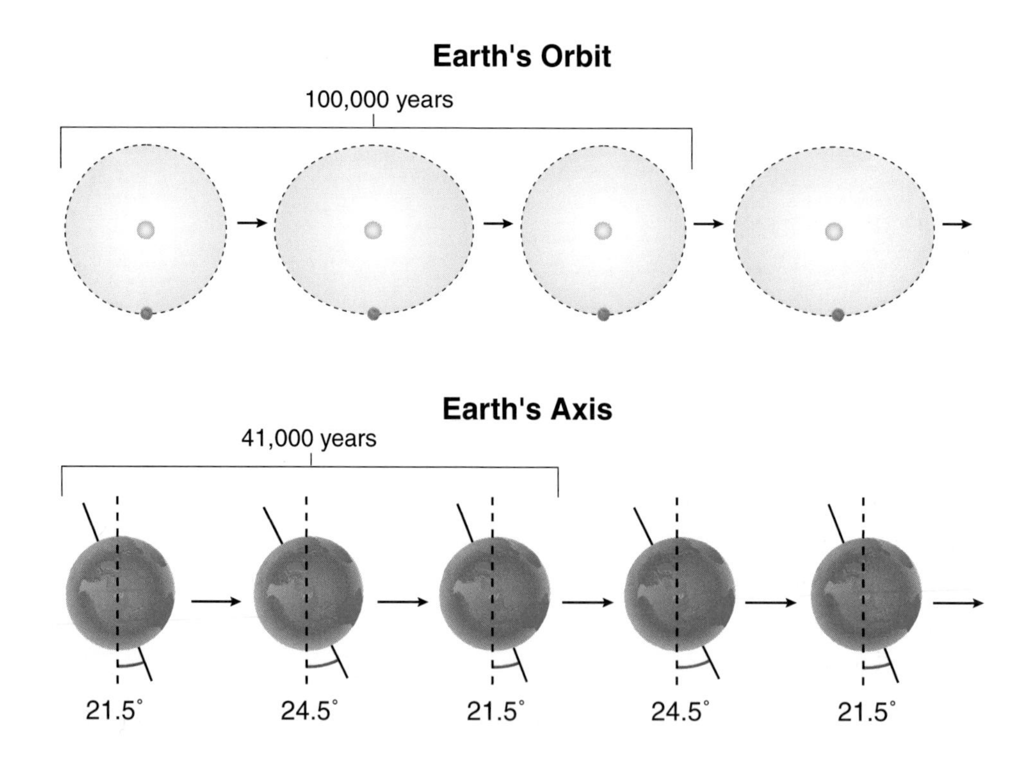

How Solar Energy Circulates Within the Earth System

Tropical areas near the equator receive much more sunlight than the polar regions. The result: places like Hawaii, Ecuador and Egypt are much warmer than Iceland and Antarctica. Actually, based on the amount of solar energy they receive, we would predict that the places closer to the pole (especially in the Northern Hemisphere where most of the land is) would be much colder than they are. However, the atmosphere and the oceans carry heat from the tropics toward the poles. Without this circulation, cities such as London, Paris, Moscow and Berlin would be much colder than they are today.

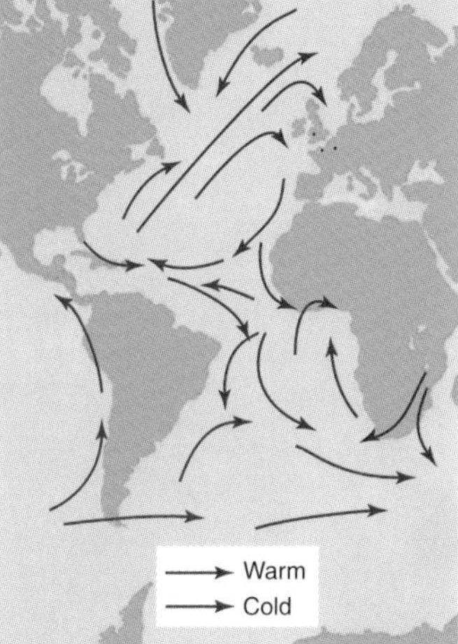

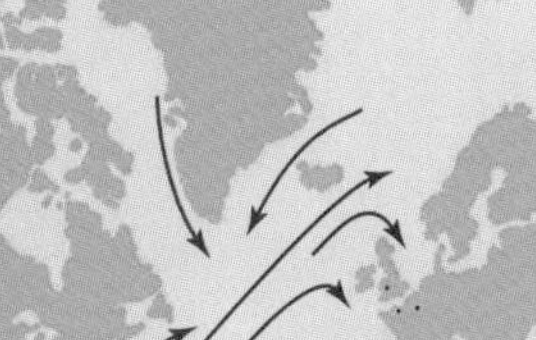

How Heat Energy Leaves the Earth System

The greenhouse effect (see pages 50-53) plays the key role here. Heat radiating from Earth's surface does not immediately leave the Earth system. Greenhouse gases in the atmosphere absorb the heat rays and send them back to Earth. As a result, the energy remains longer in the Earth system and Earth is 33 degrees C (60 degrees F) warmer than it would be without the greenhouse effect.

Water vapor and carbon dioxide are the two main natural greenhouse gases. Scientists have analyzed air bubbles trapped in the Antarctic ice sheet. The data indicate that the level of carbon dioxide in the atmosphere is strongly connected with Earth's climate. Periods of higher amounts of CO_2 correspond with warmer climates and times with lower CO_2 correspond with colder climates.

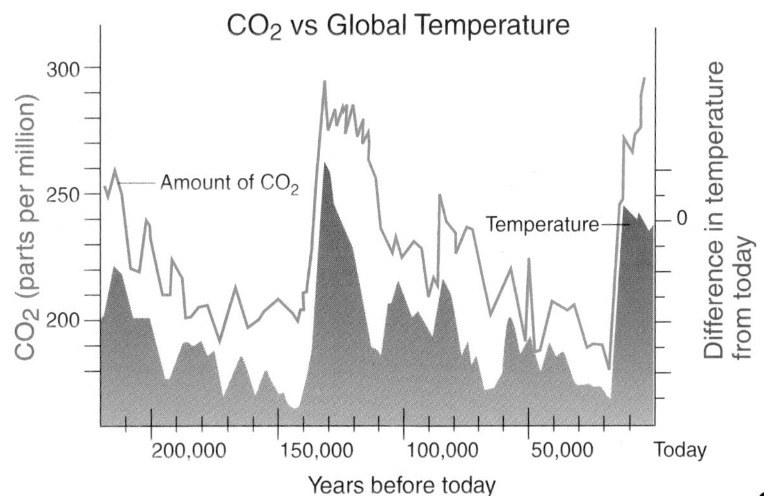

ice age or hot house?

We would like to know the answer to what sounds like a simple question. Can we expect the climate to remain the same, become colder or become hotter?

Based on the pattern for the past 160,000 years, we would predict that Earth will go deeper into an Ice Age within the next ten thousand years. However, human activities are adding at least six different greenhouse gases to the atmosphere. We might therefore predict that Earth will experience global warming. In fact, the evidence seems to indicate that this global warming has already begun. The last two decades of the twentieth century were the warmest in recorded human history.

Computer models predict that global temperatures will increase 1 to 4 degrees C within the next 100 years. That may not sound like much, but consider that the coldest and warmest periods in the past several million years involved changes of just 5 to 10 degrees C. Further, we are making our changes at an extremely rapid rate. The previous warming averaged about 1 degree C per thousand years. We may be causing temperatures to change 10 to 40 times faster.

We don't know how the Earth system will respond to these changes. Earth's climate results from a truly awesome combination of interlinked feedback loops (see pages 72-73). Some examples include:

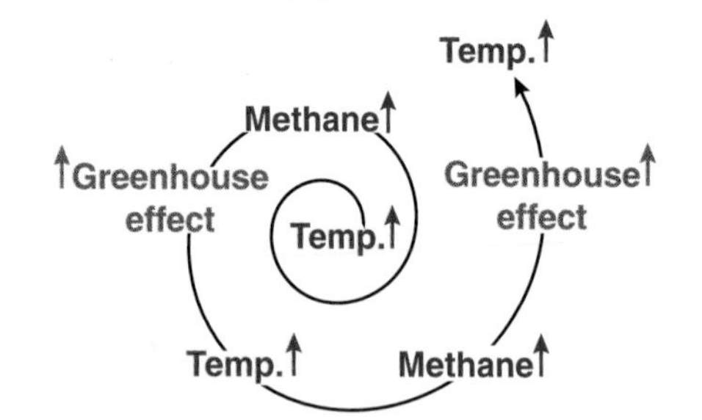

REINFORCING FEEDBACK

As the northern polar regions warms, frozen methane (a greenhouse gas) could evaporate and cause even more warming.

Temperature keeps getting higher.

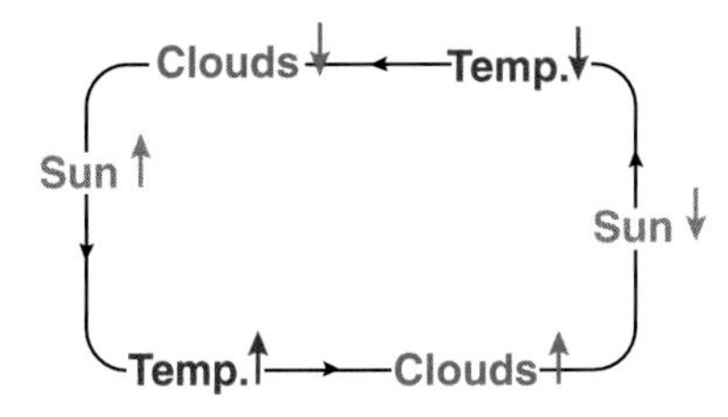

BALANCING FEEDBACK

A warmer world will have more clouds. These could make the climate colder by blocking sunlight.

Temperature stays balanced.

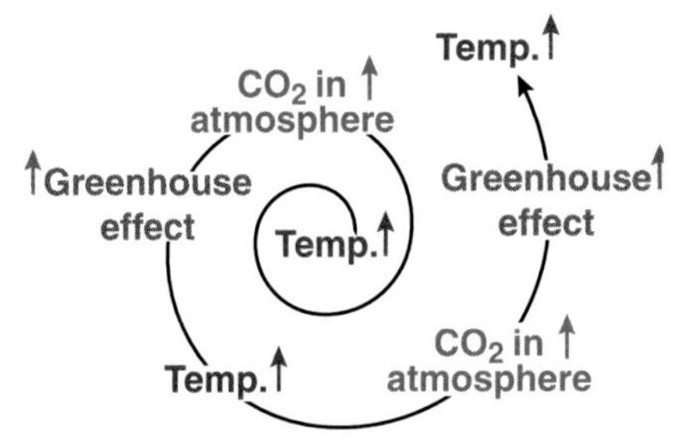

REINFORCING FEEDBACK

As the ocean warms, it may absorb less CO_2, thereby increasing the amount in the atmosphere.

Temperature keeps getting higher.

SURPRISE!!

Higher temperatures could cause changes that stop the ocean currents that carry the heat from the Equator toward the North Pole. Cooling of northern land areas could trigger a severe Ice Age.

Warm temperature causes an Ice Age!!

The take-home message is that we are conducting an uncontrolled experiment with Earth's cycles of matter, flows of energy and web of life. A warming climate could cause sea levels to rise; increases in severe weather events (tornadoes, hurricanes, floods); massive changes to agriculture; movement of diseases such as malaria into new areas; and added stress on the web of life. The social and economic costs could be enormous. Should we change? What choices do we have? Does the last chapter have any answers?

Explore Chapter 5 on the web...

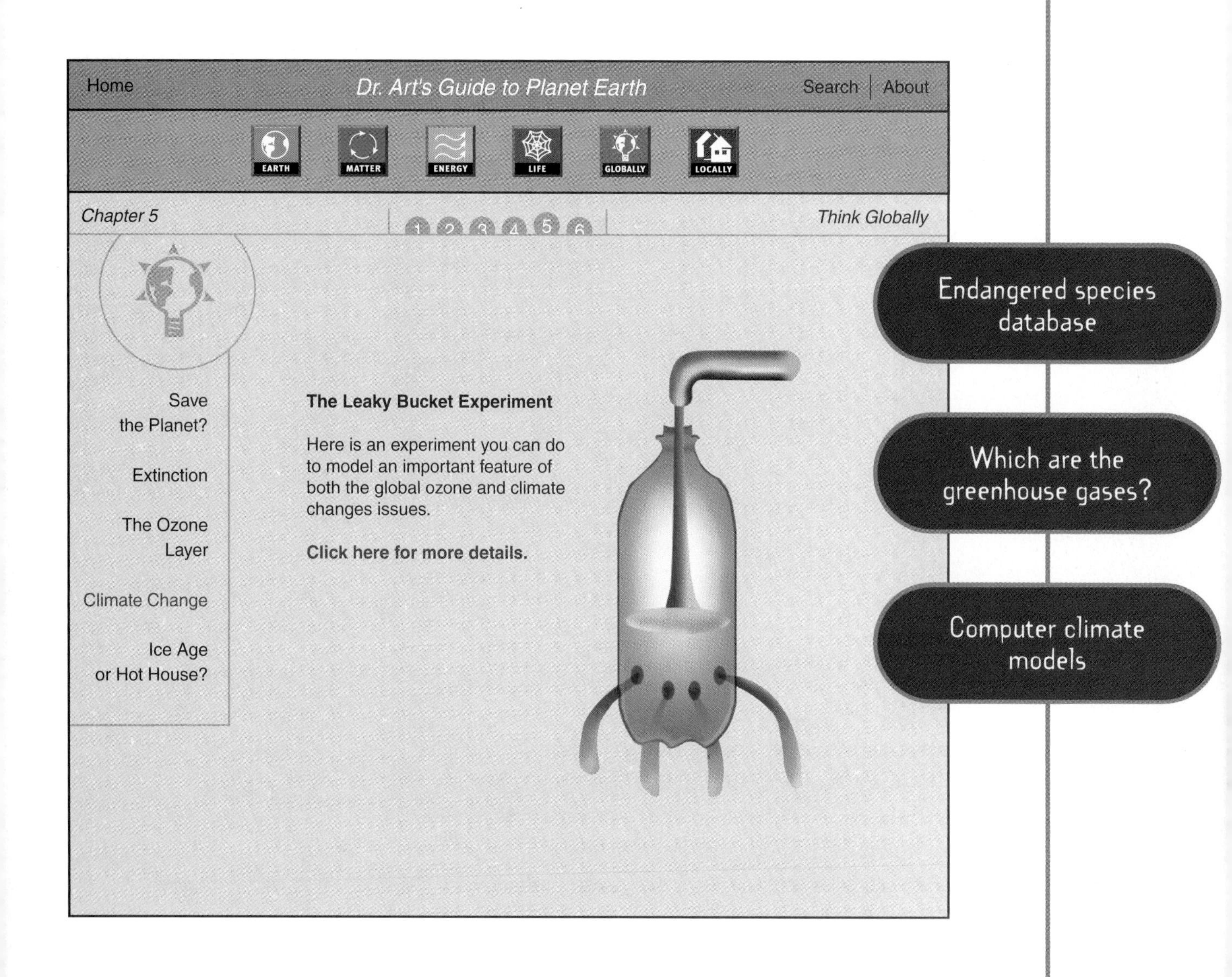

www.planetguide.net

chapter 6

act locally

Healthy Air, Water & Food

The Three R's

Local Ecosystems

What About Energy?

What Can I Do?

Making A Difference

Not The End

healthy air, water & food

How would you feel if you saw the following signs in your neighborhood? "PUBLIC HEALTH WARNING. Don't drink the water. Don't breathe the air. Don't eat the food."

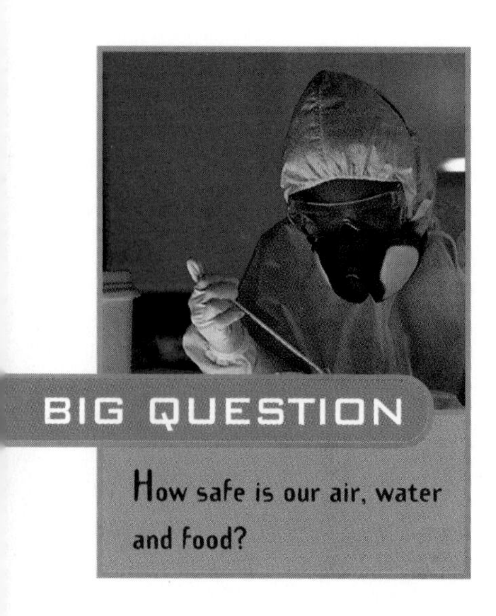

BIG QUESTION

How safe is our air, water and food?

Most of us get very concerned when we hear that our air, water, or food could make us sick rather than keep us healthy. We may hear warnings on the radio to stay indoors on certain days because the local air has high smog levels. In 1997, the United States Environmental Protection Agency reported that 107 million Americans experienced unhealthy air, mainly ozone caused by using fossil fuels.

How safe is our air, water and food? In many parts of the world, people still get very sick and even die because their water has organisms that cause diseases such as cholera. In comparison, those of us who live in the developed world generally do not have to worry that we will get very sick from germs or worms contaminating our water, air or food. Our local environmental health issues relate to chemicals that may be in our air, water, and food.

You would probably be surprised if you made a list of all the chemicals in the products we use, the fuels we consume, and the food we eat. We make and use thousands of different substances that either previously did not even exist in nature or were present in local environments in very limited amounts. These include simple elements such as lead in paints as well as complicated substances whose names are so long that we only call them by their initials.

Remember "matter cycles," one of our three Earth systems principles? Nothing simply disappears. When we put chemicals in our air, water, and land, we should not be surprised if they increase in amount. They may not fit into the existing, natural cycles of matter. Each chemical will have its own pattern of where it accumulates, how it breaks down, and how it interacts with other substances and with organisms such as humans. Concerns in the developed world about healthy air, water, and food generally arise because we have not paid enough attention to the principle that matter cycles.

the three R's

Healthy air, water and food is one local environmental issue. A second local issue concerns the vast majority of substances that we encounter that are not toxic. Even the safest things that we use create a local environmental issue that most of us deal with every day: garbage. After we have finished using something, we have to figure out where to put it.

DID YOU KNOW?

Industry moves, mines, extracts, shovels, burns, wastes, pumps and disposes of 4 million pounds of material in order to provide one average middle-class American family's needs for a year.*

Both of these local environmental issues (environmental health and garbage) result from not paying enough attention to the cycles of matter. To maintain a healthy environment, we care about the quality of the substances that we use. We want to make sure that we are not exposed to toxic chemicals as a result of how they are made, used and thrown away. In the case of garbage, we care about the quantity of stuff that we use, where it came from, and where it goes.

Try the following experiment. Carry a garbage bag with you for an entire day. Instead of throwing anything away, put it in the garbage bag. At the end of the day, weigh your garbage bag. Add 5 pounds for every gallon of gasoline that you burned during the day (calculated by dividing the miles you traveled by the fuel efficiency of miles per gallon per person). Now, multiply that weight by 40 to calculate how much solid waste you produced that day.

* *Natural Capitalism* by Paul Hawken, Amory Lovins and L. Hunter Lovins.

The gasoline helps account for the waste produced in meeting all our energy needs. But why multiply by 40? Because we do not see almost all the solid waste that we produce. Industry created about 40 pounds of solid waste in order to make each pound of stuff that we threw away. As one extreme example, it takes about 20,000 pounds of stuff to make a five-pound laptop computer.

Three R's can drastically reduce these awesome amounts of garbage that we produce.

Reduce

use less stuff. Examples include deciding you don't really need another pair of new shoes, buying products that use less packaging and that last longer, and saving energy.

Reuse

use the same stuff over and over again. Examples include canvas shopping bags, buying previously worn clothes, and fixing something rather than throwing it away.

Recycle

make new stuff from old stuff. Examples include composting, aluminum cans, and recycled paper.

Consumers can reduce garbage by practicing the Three R's and by supporting businesses that pay more attention to Earth's cycles of matter.

local ecosystems

I grew up in New York City. My environment was tall apartment buildings standing side by side with no space in between, and city streets with double-parked cars. As a child, I thought my local park was a place where they had dumped dirt on top of the concrete so grass, trees and squirrels could live there.

BIG IDEA

We cannot change today's urban environments back into wilderness.

Just 500 years ago, people's local environments looked very different. With less than one-tenth the number of people and a much higher percentage living on the land by hunting and farming, people connected much more directly with the natural world. Their local ecosystems resembled, or even were, what we call wilderness today.

We cannot change today's urban environments back into wilderness. We have permanently changed habitats; killed the local animals and vegetation; brought in our favorite animals and plants; and polluted the land, air and water. However, if we want, we can reduce the amount of new damage that we cause, and we can even begin to improve the biological health of our local environments.

Chicago's UrbanWatch shows how an urban community can investigate and help restore its local environment. Thousands of Chicago residents participate under the leadership of the Field Museum of Natural History to scientifically investigate natural areas in their local environments. Research scientists at the Field Museum and the Illinois Department of Natural Resources help decide which organisms people should observe, provide information and skills, and analyze the data gathered by the participating families, students and community groups.

People investigate city green spaces such as backyards, vacant lots, parks and golf courses. And guess what? The scientists want them to look for those "invisible," smelly and ugly organisms that we highlighted in the last chapter as playing important roles in ecosystems. UrbanWatch participants use the web (www.fmnh.org/UrbanWatch) to learn about Chicago organisms and to share the data they gather. The project aims to use this information to help scientists, policy makers and local residents to protect and nurture their local urban ecosystems.

Chicago's Urban Watch helps city dwellers identify organisms in their local ecosystems.

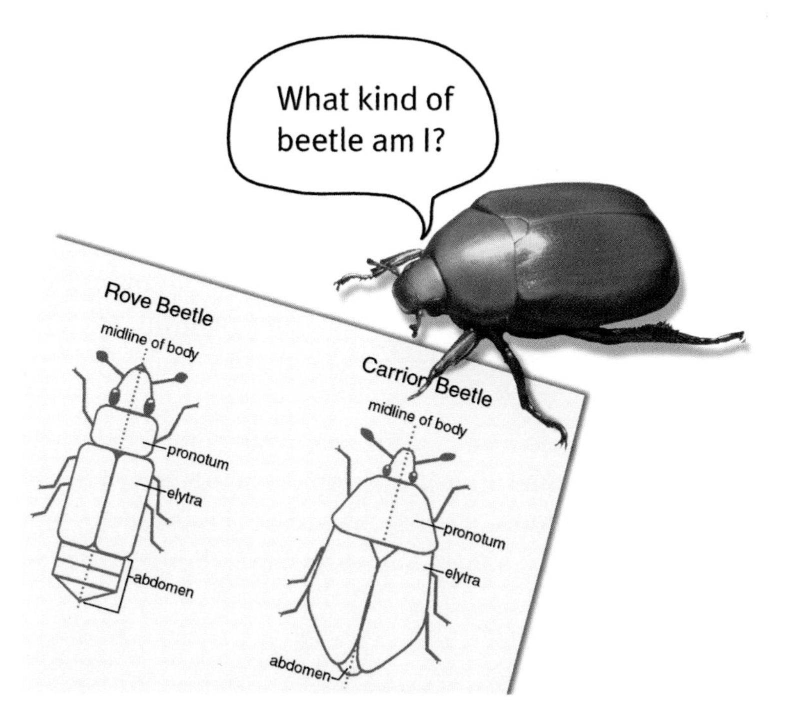

what about energy?

So far in this chapter, we have discussed matter and life. How does energy connect with local environmental issues?

Obviously, energy plays a very important role in our daily lives. We use energy when we go from place to place; heat, cool, and light our homes and businesses; grow, distribute, preserve and cook our food; and clean ourselves and our clothes. In everything that we do, we use a source of energy such as gasoline for the car, natural gas for the stove, electricity for the refrigerator, or sunlight to heat the water.

Currently, most of that energy comes from fossil fuels. Coal, oil and natural gas account for about 80% of the commercial energy consumed in the U.S. and worldwide. Whenever we burn a fossil fuel, we release carbon dioxide and thereby increase the greenhouse effect. Use of fossil fuels also causes pollution from the combustion, mining, transporting and refining processes.

These pollution and greenhouse issues all take us back to the principle that Matter Cycles. Whenever we use stuff to get energy, we have to pay attention to where that stuff came from and where it goes.

None of us wants to burn oil, coal or natural gas for its own sake. We want services such as transportation, heat, light, entertainment, etc. Can we use energy sources to get these and reduce our impacts on the cycles of matter?

Amory Lovins, a physicist and energy consultant, answers with a definite yes. The Rocky Mountain Institute, led by Amory and Hunter Lovins, claims that we can save huge amounts of energy by using the latest energy efficiency technologies in our cars, homes, businesses and industries. Their home/office in Snowmass, Colorado (where winter temperatures can reach minus 40 degrees) is so well designed that the sun provides 99% of its space and water heating. Located at over 7,000 feet elevation, the Lovins harvest bananas in December that grow inside their home/office!

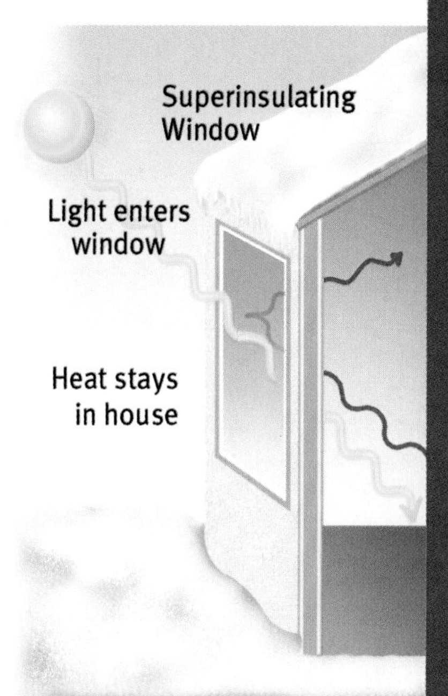

Their approach is to radically reduce the amount of energy that is needed for services such as heating or cooling. For example, they use superinsulating windows that provide lots of natural light. Unlike the usual windows, these act as a barrier preventing energy from flowing into or out of the house. In contrast, a typical home or office needs to consume fuel to make up for the heat lost through the windows in winter and for the heat brought in through the windows during the summer.

People and groups who push for strongly improving energy efficiency also often argue that society can and should meet its remaining energy needs using renewable energy sources such as solar, wind and hydropower. When we studied energy in Chapter 3, we discovered that the sun provides 15,000 times as much energy as human societies consume today. These renewable energy sources tend to cause less pollution than fossil fuels, and they generally do not increase the greenhouse effect. On the down side, renewable energy sources tend to be less convenient.

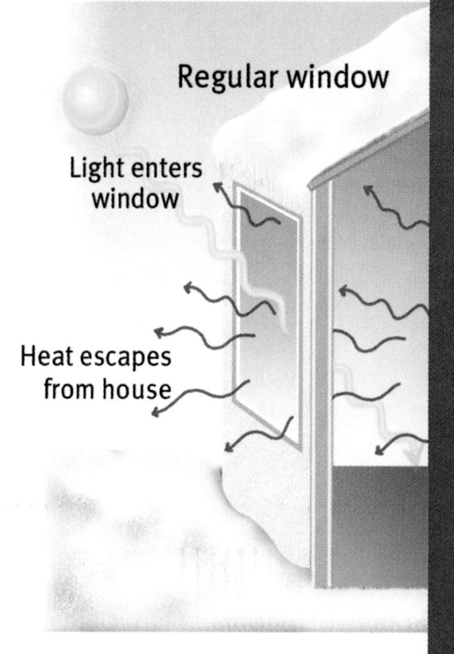

Other people and groups argue that we do not need to be as concerned with the greenhouse effect, that fossil fuels can be used more cleanly, and that switching to renewable sources will cost too much and hurt the economy. Almost everyone agrees that improving efficiency makes sense, but there is disagreement over how much energy can be saved this way. I agree with the Rocky Mountain Institute that a combination of energy efficiency and renewable energy sources can enable people in both the developed and developing nations to have a high quality of life with fewer environmental impacts than today.

what can I do?

Imagine that today is your lucky day. A very rich person says she will give you a million dollars a day for thirty days OR a quarter today, 50 cents tomorrow, a dollar on Day 3, and will continue doubling the amount for 30 days. Which would you choose?

How Much Do You Get Each Day?

Day	Amount
1	$.25
2	$.50
3	$1
4	$2
5	$4
6	$8
7	$16
8	$32
9	$64
10	$128
11	$256
12	$512
13	$1,024
14	$2,048
15	
16	
17	
18	
19	
20	$131,072
21	
22	
23	
24	
25	$4,194,304
26	
27	
28	
29	
30	

In the first choice you would get 30 million dollars total. In the second choice, even though you start at just 25 cents on Day 1, you would get 134 million dollars on the 30th day! Do the math yourself to see how this works.

We use the term "exponential growth" for this kind of explosive increase in amount. Humans today can change the way our planet operates because exponential growth has vastly increased both our population and the amount of materials that we use. Modern humans originated about 200,000 years ago. It took all our pre-history until the year 1800 before the population reached 1 billion. Then it took 130 years for that number to increase by another billion. Nowadays Earth's human population increases by 1 billion people about every 12 years.

EXPONENTIAL GROWTH OF HUMAN POPULATION

Population Total (Year)	Number of Years
1 billion (1800)	200,000 years
2 billion (1930)	130 years
3 billion (1960)	30 years
4 billion (1975)	15 years
5 billion (1987)	12 years
6 billion (1999)	12 years

Most of that population lives in developing nations, places such as China, South America, India, Africa and Indonesia. About 15% of the world population lives in the developed world in places like the U.S.A., Western Europe, and Japan. Even though they are a minority, the citizens of these countries tend to have a greater impact on the environment because of their high consumption levels and technology-based societies. A typical American uses 106 times as much commercial energy as a citizen of Bangladesh.

If everybody lived the way Americans do, the environmental impacts would be much larger than they are today. Many of us who live in the developed world understand this situation and care about the environment. Nine out of ten Americans agree that protecting the environment will require most of us to make major changes in the way we live.

DID YOU KNOW?

A typical American uses 106 times as much commercial energy as a citizen of Bangladesh.

But what can we do? We hear about recycling, choosing paper or plastic at the supermarket, turning off lights, carpooling, and selecting reusable or disposable diapers. Which of the many things that we could do would actually make the biggest difference?

A book from the Union of Concerned Scientists shows one way that we can decide. Called *The Consumer's Guide to Effective Environmental Choices*, this book analyzes American society and explains how the different things that we do in our daily lives affect the environment. I include their conclusions in *Dr. Art's Guide to Planet Earth* because the authors, Dr. Michael Brower and Dr. Warren Leon, use science-based, systems thinking. They use the systems approach that we described back in Chapter 1. Brower and Leon analyze the different parts of our consumption system, study how these parts connect with each other, and include how they are part of the larger environment.

The bar graph summarizes their conclusions. First, they decided to focus on four major kinds of environmental impacts. Two of these are global environmental issues that we discussed in the previous chapter (global warming = global climate change; habitat alteration = extinction/loss of biodiversity). The other two environmental impacts are major local environmental issues (air pollution and water pollution). They did not include our third global environmental issues (ozone layer). Since CFCs are no longer manufactured, our actions as consumers cannot affect this issue as much.

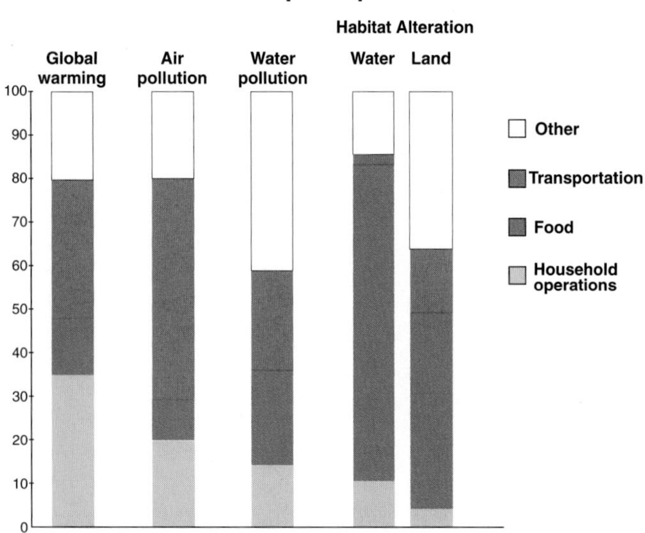

According to this study, three different kinds of activities account for most of our environmental impacts as consumers. These are transportation, food, and household operations. In other words, we should pay the most attention to how we go from place to place, what we eat, and how we keep our homes operating (especially heating, cooling, and lighting). These three kinds of activities account for about 75% of our consumer impacts on global warming, air pollution, water pollution, and alteration of habitat.

Transportation accounts for 32% of our consumer impacts on global warming and 51% of our impacts on toxic air pollution. Most of this comes from our favorite tool and toy, the personal car/light truck. A big part of the solution is to drive less, and use vehicles that get more miles to the gallon and release the least pollutants. If we drive a gas guzzler to pick up a gallon of milk at the market, it does not matter whether we use paper, plastic or no bag. We created much bigger environmental impacts driving there and back.

Food has enormous environmental impacts, particularly in the areas of water pollution and habitat alteration. Growing food and grazing livestock occupy 60% of the U.S. land area. Fertilizers, pesticides, animal wastes and erosion

all affect water quality. Brower and Leon recommend that we reduce the amount of red meat that we eat. Red meat has much greater environmental impacts than poultry or grain. Compared to pasta, red meat causes 18 times as much water pollution and 20 times as much impact on land use. They also recommend that we eat organic grains, vegetables and fruit. Organic farming produces less water pollution because it does not use synthetic fertilizers and pesticides.

Household operations is the third large category of consumer impacts on the environment. In many of our homes, we burn fossil fuels to heat our living space and water. We use electricity for these purposes as well as for lighting, refrigeration and appliances such as the TV, computer and stereo. More than half the electricity in most countries, including the U.S., comes from burning fossil fuels, especially coal. And most of our homes use these energy sources very inefficiently.

Brower and Leon summarize Eleven Priority Actions for American Consumers:

11 PRIORITY ACTIONS FOR CONSUMERS		
Transportation	**Food**	**Household Operations**
1. Choose a place to live that reduces the need to drive. 2. Think twice before purchasing another car. 3. Choose a fuel-efficient, low-polluting car. 4. Set concrete goals for reducing your travel. 5. Whenever practical, walk, bicycle, or take public transportation.	6. Eat less meat. 7. Buy certified organic produce.	8. Choose your home carefully. 9. Reduce the environmental costs of heating and hot water. 10. Install efficient lighting and appliances. 11. Choose an electricity supplier offering renewable energy.

Each of us has to decide how important we think these global and local environmental issues are, and what we are willing to do about them. I have shared the study and conclusions from the Union of Concerned Scientists as one systems-based approach that can help a person choose which actions can have the most beneficial effects. The next section features two groups that use systems thinking to make an environmental difference in their communities.

making a difference

Back in 1994, Ray Anderson prepared a speech that changed his life. More than twenty years earlier, he had founded Interface, Inc., a company that is the world's largest producer of commercial floor coverings, making and selling more than 40% of all the carpet tiles used on Earth. Ray's speech helped make Interface become much more exciting than a company that makes rugs for businesses.

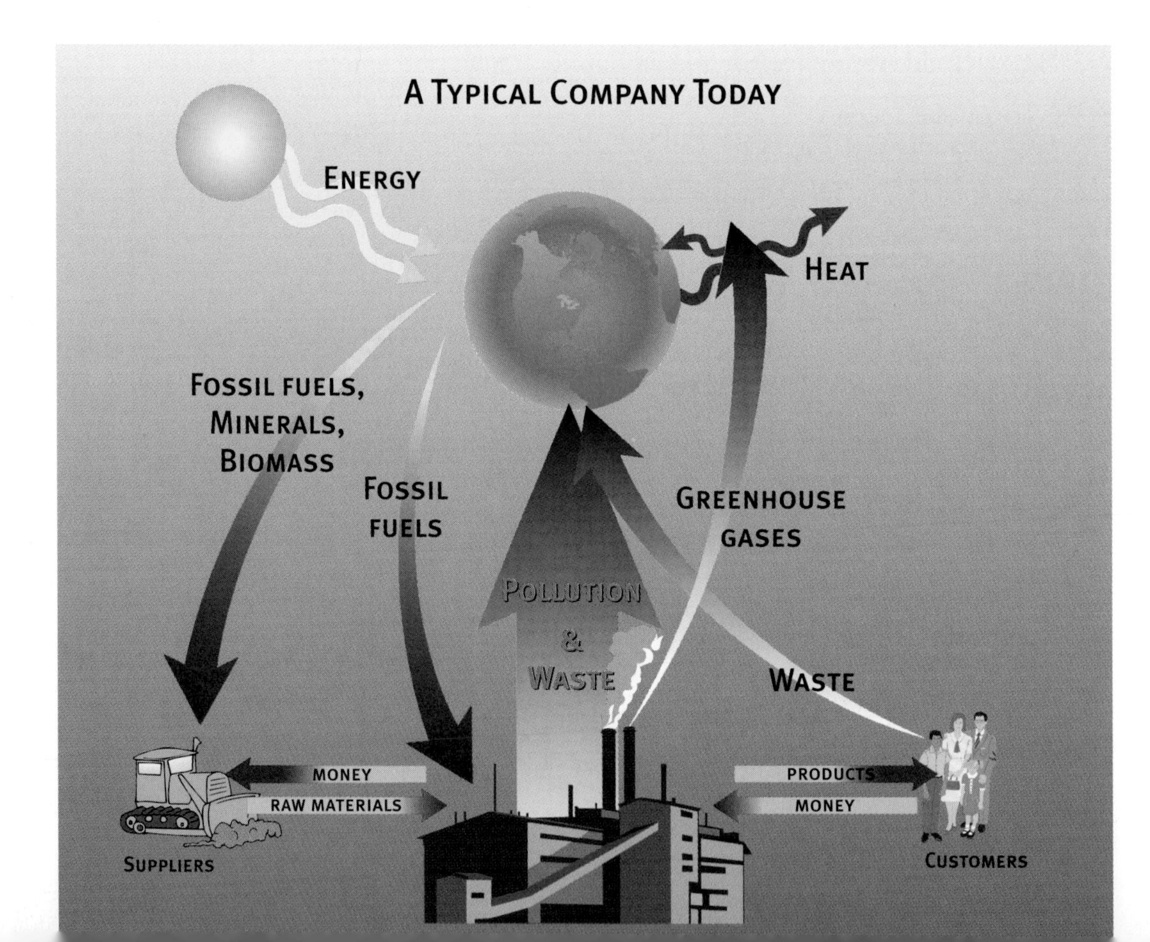

Ray had been asked to talk about his company's environmental vision. In preparing his speech, he realized that they did not have one. As he read and explored, he decided to change Interface from a company that was damaging the environment into one that is restoring it. And to keep making carpets while increasing sales and profits.

Since then, Interface has used systems thinking to eliminate waste and pollution. They calculated how much stuff they were taking from the Earth in order to make their products, and discovered that they were using 1.2 billion pounds of Earth materials, mostly fossil fuels. Ray Anderson says it made him want to throw up.

Five years later, Interface had reduced waste by about 50% and saved a lot of money in the process. The company uses at least 5 R's, beginning with reduce, reuse, and recycle, and adding redesign and renewable energy. Their newest product will create zero waste and will be produced using solar power rather than fossil fuels.

A Sustainable Company

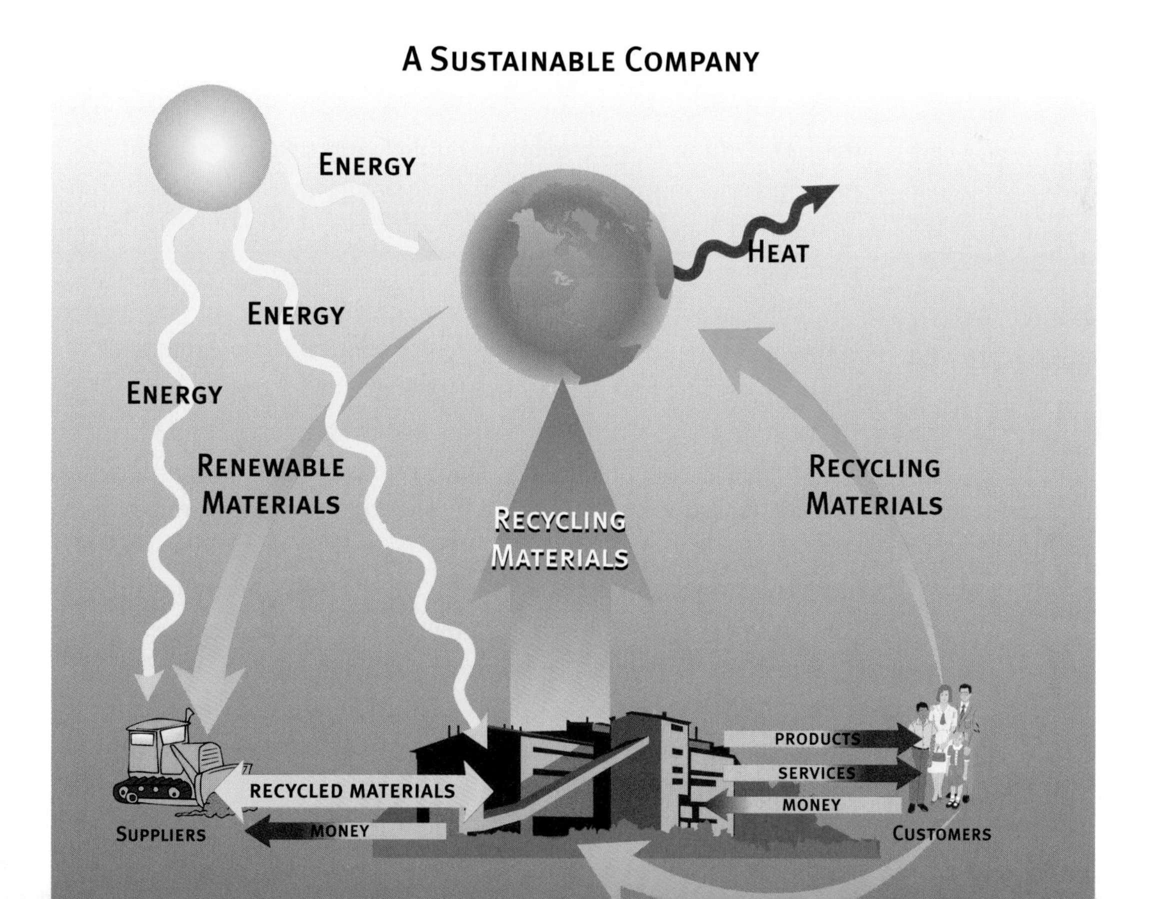

Another redesign feature is that a customer can purchase rug services rather than rugs. In the old way, a customer buys a rug, and then throws it out when it needs replacing. In Ray's scheme, the customer buys a lease that provides constant, high quality floor covering. When a rug section needs replacing, Interface takes out the worn section, replaces it, and completely recycles it. Zero waste and a better deal for the customer. To quote Ray Anderson, Interface does well by doing good.

TreePeople in Los Angeles provides another example of a group that does well by doing good. Founded in 1973, this non-profit organization supports Los Angeles residents in improving the neighborhoods in which they live, work and play. Started by a group of teenagers, Treepeople now has planted over a million and a half trees in Los Angeles and has one of California's largest environmental educational programs.

State and federal laws require the Los Angeles county government to meet a variety of environmental standards (such as clean water, reduced trash and clean air). TreePeople developed a program for the L.A. County Department of Public Works to educate and support teens in improving their local environments. In this Generation Earth program, they have involved 800,000 high school students in just two years.

Looking at the urban system, TreePeople selected teens as a target audience because they can strongly influence their families and friends, and because the choices they routinely make every day add up to large environmental impacts. Generation Earth helps teens understand the environmental systems in which they live and what they can do to protect them.

The campus water project is one example. Generation Earth teaches the water cycle from the points of view of the planet and of Los Angeles. Students learn where their water comes from and where it goes. Most of us would think about the water that flows in our faucet's and out of our homes into the public sanitary sewer system. But another important flow of

city water occurs outside this system. This is the water that comes from clouds or garden hoses, and ends up running down our streets, into the storm drain system and flowing into the ocean.

This urban runoff picks up pollution from car oil, cigarette butts and other city trash. Unlike the sanitary sewer system, urban runoff is not treated before it flows into the ocean where it can pollute water and beaches. Generation Earth students learn these water issues and discover how they can prevent stormwater pollution. They analyze how their school fits within this water system and then engage in action projects to minimize the school's harmful environmental impacts. Projects such as Generation Earth not only solve problems today, but they also help create future citizens who have environmental understanding and the skills to make a difference.

not the end

Each of us has strong beliefs and values that guide our actions. With respect to Earth, one of my values is to leave our planet in at least as good a condition as I have enjoyed. What are your values?

I hope this book has given you a simple framework for understanding how our planet works. It should also have shown you how complicated the Earth system is, how much we don't know about such basic issues as the number of different species or what the climate will be in the next fifty years. If I don't understand something and my life depends on it, I tend to be cautious about messing with it. That makes me conclude that we should:

Maintain the current balance in matter cycles

Avoid interfering with Earth's energy flows

Preserve the web of life

You could have a very different value system and reach different conclusions. I simply ask two things. First, that you try to understand planet Earth as a system. Second, if you decide that you want to help protect the Earth system, then use systems thinking to choose what you do. Every choice we make has both good and bad features. Systems thinking can help you balance those risks and benefits. It can also help you identify the places in the system where you can have the most effect.

We can do many things in our daily lives that make a huge difference. Brower and Leon identified eleven priority actions in the areas of transportation, food and household operations. I also think that we need to help make changes in our society that make it easier for people and businesses to act in environmentally responsible ways. As one example, it is usually much more convenient for people to drive their cars than to use public transportation. Our society has chosen to put more money into expanding highways than into developing attractive and efficient mass transit. To have the most effect, we need to combine how we vote in our daily actions with how we vote in our annual elections.

I hope this book has given you a sense of hope rather than despair. We cannot destroy life on Earth. Our human ingenuity has brought us to this unique moment in our development as a species where we can change the way our planet works. Each of us in our daily lives influences this global system and our local environment. We do not know what will happen.
I believe that the more we preserve Earth's cycles of matter, flows of energy and web of life, the more likely are our chances of preserving a hospitable planet for ourselves, our descendants and all Earth's creatures.

NOT THE END

GLINDEX
(glossary + index)

Biodiversity - the number and kinds of Earth's organisms. We don't know how many species there are, how many are going extinct today or the consequences for Earth's web of life **p. 66-67, 74-75, 80-85.**

Cycles - a repeating pattern such as the seasons of the year. See Chapter 2 for the rock cycle **(p. 20-25)**, the water cycle **(p. 26-33)** and the carbon cycle **(p. 34-41).**

Earth - a recycling planet powered by the flow of solar energy that supports a networked web of life. Introduced in Chapter 1 **p. 2-17.**

Ecosystem - the organisms that live in a particular place, and how they interact with each other and with their local environment. All ecosystems have a similar pattern of organization **p. 68-75, 83-85, 104-105.**

Electromagnetic Spectrum - the very wide range from radio waves (long wavelengths) to visible light to X-rays and cosmic rays (short wavelength); an important scientific concept that explains colors and the greenhouse effect **p. 50-53.**

Energy - a simple word whose scientific definition sounds like a riddle **(p. 44)**. Energy flows into, through and out of the Earth system **p. 12-13, 43-58.**

Energy Efficiency - how much value we get from each amount of fuel that we consume. Today's societies tend to use energy very inefficiently. We can consume much less energy and still obtain the same services and comfort **p. 106-107, 110-111.**

Feedback Loops - one way that parts of a system influence each other. Balanced feedback loops keep systems stable. Reinforcing feedback loops can cause radical change in ecosystems **p. 71-73, 96-97.**

Fossil Fuels - coal, oil and natural gas. Formed from ancient living organisms, they are a carbon reservoir. Our reliance on fossil fuels as an energy source is increasing the amount of carbon dioxide in the atmosphere **p. 37-41, 106-107, 111, 113.**

Global Climate - Earth's pattern of temperature and precipitation. Earth's climate has changed throughout its history. We don't know how much or how fast our activities will change the global climate **p. 57, 79, 82, 92-97.**

Greenhouse Effect - specific gases in the atmosphere (especially water and carbon dioxide) absorb the heat that is leaving the planet and thereby keeps Earth warmer. By burning fossil fuels and forests, we may have too much of a good thing **p. 50-53, 56-57, 95-97, 107, 110.**

Matter - the stuff that is in our world. On Earth, we encounter it in solid, liquid and gas forms. Don't be surprised if you walk around muttering "matter cycles, matter cycles, matter cycles" after reading **p. 10-11 and Chapter 2 p. 19-42.**

Molecule - matter is made of atoms. A molecule is two or more atoms joined together. The smallest piece of water is a water molecule, which consists of two hydrogen atoms joined to one oxygen atom **p. 28.**

Ozone - a form of oxygen containing three instead of two atoms joined together. Ozone in the upper atmosphere protects life from the sun's ultraviolet radiation. CFCs and other chemicals can damage this ozone layer. In the lower atmosphere, ozone is a pollutant **p. 79, 86-91, 100.**

Photosynthesis - how Earth's plant life captures solar energy and packages it in sugars; a part of Earth's carbon cycle that removes carbon dioxide from the atmosphere **p. 62-65.**

Renewable - sources of energy and materials that can be naturally re-supplied so humans can take advantage of them without using them up. Examples are sunlight, wind power and wood **p. 107, 113.**

Respiration - how organisms release chemical energy from sugars by combining them with oxygen; a part of Earth's carbon cycle that releases carbon dioxide into the atmosphere **p. 62-65.**

System - a whole that is more than the sum of its parts. You, a car and a sandwich are all examples of systems **p. 4-7.**

Systems Thinking - a way to understand our world by investigating its systems. Dr. Art recommends asking three systems questions **p. 4-9.**

You're holding the book.

On the web site you can...

- *find animations and experiments*
- *ask questions and get answers*
- *learn about Dr. Art's show and when it is happening.*

www.planetguide.net

ABOUT THE AUTHOR

Dr. Art Sussman received his Ph.D. in Biochemistry from Princeton University. He performed scientific research at Oxford University, Harvard Medical School and the University of California at San Francisco. For the past 25 years, he has helped the general public, teachers and students understand science, especially as it affects them in their daily lives. Dr. Art works at WestEd (one of ten regional educational laboratories created by Congress) to improve science and environmental education at the local, state and national levels. He uses innovative ways to show that science is understandable, interesting, relevant and fun.

Special Offer for Group Purchases

We want to encourage groups to use *Dr. Art's Guide to Planet Earth* as part of their work in improving their communities, educating teachers, teaching students and using sustainable business practices. Examples include:

- **Universities or projects that educate future teachers**
- **Businesses that want to inform their employees, clients and/or customers**
- **Students or community groups involved in environmental projects**
- **Schools that teach Earth systems science**

Please send an e-mail to **planetguide@wested.org**. Tell us how many copies of the book you want and provide a brief description of your situation/project.. We will offer a group purchase discount.